WATCH REPAIRING

Cleaning, and Adjusting

F. J. GARRARD

Arlington Book Company
1988

This edition, published in 1988 by Arlington Book Company, is an unabridged republication of the work originally published by the Norman W. Henley Publishing Company, 2, West 45th Street, New York, N.Y. U.S.A. in 1930 under the title "Watch Repairing, Cleaning, and Adjusting, A Practical Handbook dealing with the Materials and Tools Used, and the Methods of Repairing, Cleaning, Altering, and Adjusting all kinds of English and Foreign Watches, Repeaters, Chronographs, and Marine Chronometers" by F. J. Garrard.

Printed in the United States of America by

Arlington Book Company
P.O. Box 327
Arlington, Virginia 22210-0327 U.S.A.

ISBN 0-930163-17-6

WATCH REPAIRING, CLEANING, AND ADJUSTING

A PRACTICAL HANDBOOK

DEALING WITH

THE MATERIALS AND TOOLS USED, AND THE METHODS OF REPAIRING, CLEANING, ALTERING, AND ADJUSTING ALL KINDS OF ENGLISH AND FOREIGN WATCHES, REPEATERS, CHRONOGRAPHS, AND MARINE CHRONOMETERS

BY

F. J. GARRARD

SPRINGER AND ADJUSTER OF MARINE CHRONOMETERS AND DECK WATCHES FOR THE ADMIRALTY

AUTHOR OF "LATHES FOR WATCHWORK, AND HOW TO USE THEM"

WITH OVER TWO HUNDRED ILLUSTRATIONS SPECIALLY DRAWN AND ENGRAVED FOR THE WORK

Ninth Impression

NEW YORK

THE NORMAN W. HENLEY PUBLISHING COMPANY

2, WEST 45TH STREET

1930

PREFACE.

IN this volume the aim of the Author, who has himself served an apprenticeship to the trade of Watchmaking, and worked at the watchmaker's bench for the best years of his life, is to give in plain language practical directions for carrying out the various operations involved in Repairing and Cleaning Watches.

He has attempted to cover the whole ground of Watch Repairing, from the simplest operations to the most complex and difficult ; and it is hoped that the book will prove of service alike to the workman, the apprentice, and the amateur. To the first-mentioned, the chapters on Springing and Adjusting and on Complicated Watches are more particularly addressed. These chapters, in particular, contain information which it is believed cannot be found elsewhere, and will, the Author feels assured, be welcome even to those who are already adepts in the art of Watch Repairing.

The illustrations are a special feature, being all original; and the Author having tried to avoid the tendency which so many technical books have of degenerating into mere catalogues of tools and appliances, he may confidently say of the present work that it at least has the merit of being *fresh* from cover to cover.

F. J. GARRARD.

CONTENTS.

CHAPTER I.—INTRODUCTION.

PAGES

Description of a Watch Movement—Motive Power—The Train —Escapement and Balance—Motion Work. (Figs. 1–5) . 1–7

CHAPTER II.—THE MATERIALS USED IN THE CONSTRUCTION AND REPAIR OF WATCHES.

Steel, Softening, Hardening, Tempering—Brass, Softening, Hardening, Tempering—Gold—Silver—German Silver—Platinum—Palladium—Aluminium Bronze—Invar—Redstuff—Diamantine—Diamond Powder—Oilstone Dust—Emery—Water-of-Ayr Stone—Chalk—Benzine—Petrol—Oil —Methylated Spirit—Turpentine—Acids—Mercury—Pegwood—Pith—Tissue-paper—Shellac 8–18

CHAPTER III.—WORKSHOP, TOOLS, &c.

Heating the Workshop—Light—Work-board—Tools—Eyeglasses. (Figs. 6–26) 19–29

CHAPTER IV.—THE USE OF TOOLS.

Flat Filing—Pin Filing—The Use of Files—Drilling—Broaching —Cutting Screw Threads. (Figs. 27–31) 30–35

CHAPTER V.—TURNING.

PAGES

Using the Turns—Gravers—Methods of Cutting—Turning for Practice—The Watch Lathe—Its Appliances and Chucks—Methods of Driving. (Figs. 32–53) 36–47

CHAPTER VI.—MAKING SMALL TOOLS.

Drills — Taps — Punches — Screwdrivers — Soft Soldering — Countersinks—Chamfering Tools—Joint Pusher—Oiler—Blueing Slip — Hard Soldering — Brass Tweezers. (Figs. 54–61) 48–53

CHAPTER VII.—CLEANING WATCHES.

Why Watches want Cleaning—Key-wind Geneva Watches—Taking to Pieces—Cleaning—Putting together—Endshakes Oiling—Keyless Watches—Taking out of Cases—English Watches—Verge Watches—English Cylinder and Duplex Watches—Pocket Chronometers—English Keyless Watches—American Keyless Watches—General Remarks on Cleaning Watches. (Figs. 62–71) 54–69

CHAPTER VIII.—BARRELS, FUSEES, MAIN-SPRINGS, AND CHAINS.

Broken Mainsprings — Barrels, Repairing — New Barrels — Mending Chains — Fusee Clickwork — Stopwork — Safety Pinions—Going-barrel Clickwork. (Figs. 72–86) . . 70–81

CHAPTER IX.—DEPTHS, TRAIN WHEELS, &c.

Depths—Correcting Bad Depths—Worn Pivots—Polishing Pivots—Turning Pinions — Turning Pivots — Cement Chucks — Bushing Pivot Holes—Fitting New Wheel Teeth—New Pivots — Jewel Holes — Jewelling — Frames, Screws, &c. (Figs. 87–119) 82–99

CHAPTER X.—ESCAPEMENTS.

Lever Escapement—Faults of—Turning New Balance Staffs—Pivots, Repairing—Cylinder Escapement—New Cylinders—Corrections—Verge Escapement — Duplex Escapement — Chronometer Escapement — Other Escapements. (Figs. 120–146) 100–128

CHAPTER XI.—BALANCES AND HAIRSPRINGS, ADJUSTING AND TIMING.

PAGES

Balances—Temperature Error—Compensation Balance—Hairsprings—Fitting—Counting—Setting True—Breguet Springs—Forms of—Overcoils, Making—Adjusting for Temperature—Timing in Positions—Isochronism—Repairing Hairsprings—Curb Pins. (Figs. 147–163) 129–147

CHAPTER XII.—MOTION WORK, HANDS, AND DIALS.

Motion Work—Fitting Cannon Pinions—Trains—Tightening Hands—Fitting Hands—Set-hand Arbors—Enamel Dials—Gold and Silver Dials—Up-and-down Mechanism. (Figs. 164, 165) 148–154

CHAPTER XIII.—CASES

Parts of a Case—Tightening Snaps—Fitting Glasses—Taking out Bruises — Joints — Bows—Raising Domes—Erasing Monograms, &c. (Figs. 166–173) 155–160

CHAPTER XIV.—KEYLESS WORK.

Rocking Bars—Shifting-sleeve—Pendant-set, Patek Phillippe's—Winding Wheels—Fusee Keyless Work. (Figs. 174–180) 161–169

CHAPTER XV.—CAUSES OF STOPPAGE OF WATCHES.

Small Faults—Magnetism—Karrusels—Bad Mainsprings. (Figs. 181, 182) 170–175

CHAPTER XVI.—CONVERSIONS AND ALTERATIONS IN MOVEMENTS.

Making an Escapement—Three-quarter Plate Conversions—Verge Conversions—Keyless Conversions—Improving and Renovating Watches. (Figs. 183–190) 176–181

CHAPTER XVII.—MARINE CHRONOMETERS.

PAGES

Taking to Pieces—Cleaning—Adjusting and Rating—New Pivots and Repairing. (Figs. 191–193) 182–185

CHAPTER XVIII.—REPEATING WATCHES, &c.

Repeating Mechanism — Hour Repeating Work — Quarter Mechanism—Minute Repeating Work—Hammers—Action of Repeating Work—Old Forms—Speed of Repeating—Cleaning Repeaters—Old Repeaters—Clock Watches—Alarm and Musical Watches. (Figs. 194–201) 186–196

CHAPTER XIX.—CHRONOGRAPHS, CALENDARS, &c.

Stop Watches—Independent Centre Seconds—Chronographs—Starting Mechanism—Fly-back Work—Driving Work—Minute Recorders — Double Fly-back — Cleaning Chronographs—Simple Calendars—Perpetual Calendars—Arrangement of Complicated Watches—Double Time and Sidereal Watches. (Figs. 202–210) 197–208

WATCH REPAIRING.

CHAPTER I.

INTRODUCTION.

Description of a Watch Movement.

It is not proposed in this book to enter into the question of the history of the watch, nor to discuss who invented this or that portion of it, but simply to take the modern watch as it is, and describe as clearly as possible how it works and how to repair and keep it in order.

It will, perhaps, be well first to describe, in general terms,

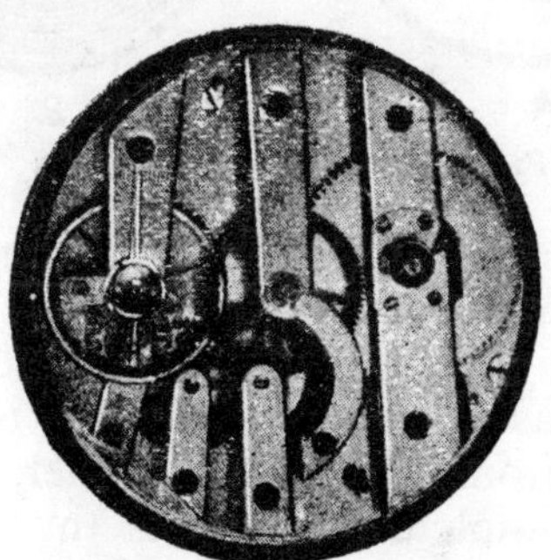

Fig. 1.—Geneva Bar Movement.

the mechanism of a watch, and for this purpose a Geneva "bar" movement will be used as an illustration. Fig. 1 shows such a movement. The term "movement," it may be explained, is applied to the *works* of a watch as distinguished from the case.

This particular movement is chosen, as its "bar" construction enables all the wheelwork to be seen. The mechanism of this movement may be divided into four portions. *First*, the motive power; *second*, a train of wheels to transmit the power; *third*, an escapement and balance to control the power; and *fourth*, motion work and hands to record the revolutions of the train wheels upon the dial.

The Motive Power.—This, in all watches, is a mainspring. A mainspring is a thin and flat strip of steel, hardened and tempered to give the maximum of strength and elasticity. It is coiled up around a steel centre arbor, to which its eye is hooked, and enclosed in a box or "barrel," to the inside of which its outer end is attached. If a barrel, containing such a spring, be held firmly, while the centre arbor is turned round,

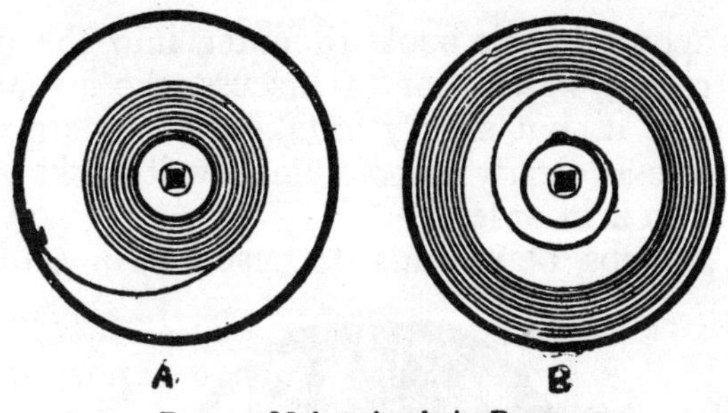

FIG. 2.—Mainspring in its Barrel.

coiling up the spring tightly around it, until the outer end pulls hard at its attachment, as at A (Fig. 2), the barrel, when released, will revolve in the direction that the spring pulls it, until the spring has unwound itself and is prevented by the containing barrel from unwinding further, as at B (Fig. 2). The number of complete revolutions thus made by a watch barrel with an average mainspring is five, and the number of revolutions used in driving the watch for twenty-four hours is generally three, thus leaving two to spare.

There are three principal methods of making a mainspring drive a watch. The first method, and the one adopted in the movement illustrated in Fig. 1, is to make the barrel into a toothed wheel by cutting teeth around its circumference. The barrel then becomes the first or "great wheel" of the watch train. This is termed a "going barrel." In a watch with this arrangement the barrel arbor is squared, and to wind the watch a key is placed upon it, and it is turned round three or four complete revolutions, being held by "clickwork." During the going of the watch, the barrel arbor is stationary and the barrel turns round, hence the term "going-barrel." "Clickwork" is the name given by watchmakers to an arrangement of a ratchet and pawl, the latter, in watches, being termed a "click." Fig. 3 shows a clickwork arrangement. In the figure A is the ratchet, B the click, and C the click-spring. In many watches the click and spring are in one piece, as in Fig. 1, but the action remains the same. In the second method the barrel is stationary while the arbor revolves, carrying with it a separate toothed wheel. In some watches on this plan the barrel is turned round in the act of winding the watch, and in others it is merely a sink recessed out in the solid watch plate, and, of course, a fixture. This method of driving is used mainly in American watches. The third method is for the barrel to be merely a drum, driving the watch by means of a chain wound upon it. This indirect and somewhat unsatisfactory method was adopted in order to equalize the force or pull of the spring. When a mainspring is fully wound up it exerts its maximum force. As it unwinds the force becomes gradually less and less, until it is zero. In the old watches the force of the mainspring directly affected the timekeeping of the watch, hence it was necessary to introduce some arrangement to equalize it. The arrangement adopted is shown in Fig. 4. A is the barrel containing the

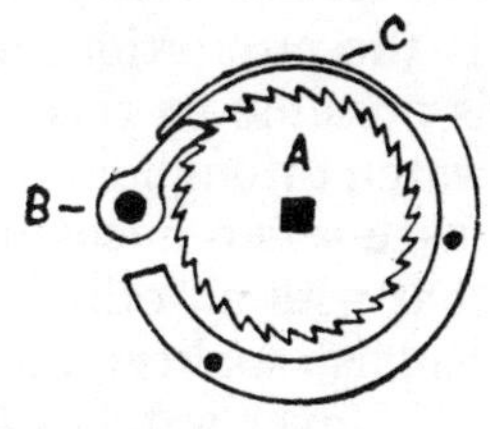

Fig. 3.—Winding Clickwork.

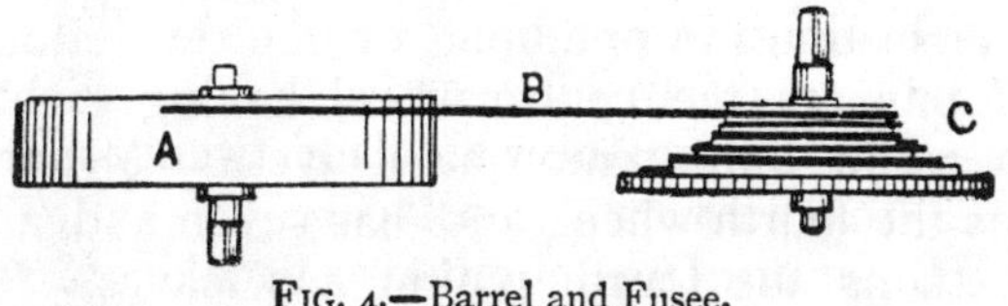

Fig. 4.—Barrel and Fusee.

mainspring, B is the chain, C is the "fusee." The fusee is a cone-shaped pulley having a continuous spiral groove cut upon it. The chain runs in this groove. When the spring is wound up the chain is on the fusee and pulls at its smallest diameter, thus exerting but a small leverage upon the fusee, to which is attached the first or main wheel of the watch train. Fig. 4 shows the arrangement when wound up. As it unwinds the barrel revolves and unwinds the chain from the fusee, coiling it up on itself. During this process the chain gets lower and lower upon the fusee body until, when nearly run down, it pulls upon the largest diameter of the cone, thus giving the diminished force of the mainspring an advantage in leverage. If the proportions of the cone are suited to the mainspring, it is possible by this means to have a constant force driving the watch throughout the twenty-four hours. With verge watches the fusee was a necessity. Though foreign makers quickly found that with all other kinds of watches the fusee was unnecessary, English makers, almost without exception, continued using it with lever watches for many years, and some use it now. In marine chronometers it is still in use.

The Train.—The mainspring thus, either directly or indirectly, drives the main wheel, which is the first wheel of the watch train. This wheel, in an average watch, turns once in eight hours. It gears into the centre pinion of the watch, causing the latter to revolve eight times to once of the main wheel, and thus turn once in one hour. This is effected by the main wheel having eight times as many teeth as the centre pinion has leaves. The centre pinion, as its name implies, occupies the centre of the watch, and its axis or "arbor" projects through the dial, and has the minute hand affixed to it. Upon the same arbor, with the centre pinion, is the centre wheel, the second wheel of the train, centre wheel and pinion forming one and revolving together. In the same way, the centre wheel drives the third wheel and pinion, and causes the latter to revolve eight times in one hour, or one revolution in seven and a half minutes, the centre wheel having eight times as many teeth as the third pinion has leaves. The third wheel, again, drives the fourth wheel, and has seven and a half times as many teeth as the fourth pinion has leaves. The fourth wheel and pinion therefore perform one revolution in one minute. A prolongation of one pivot of the fourth pinion projects through the dial and carries the seconds hand. The fourth wheel, in its turn, drives the scape pinion and wheel, causing the latter to perform ten revolutions in one minute. This completes the watch train and brings us to the escapement. In Fig. 1 all these wheels are visible, and the above explanation can be followed by reference to them. The wheels in most watches are of hard brass; a few have German silver or nickel wheels, and a few have wheels of a special alloy combining lightness and strength, such as aluminium-bronze. The pinions which they drive are of fine quality steel, hardened and tempered. The axis of a wheel is, in watchwork, called its "arbor." The pinions of the train wheels are in one piece with their arbors. Upon the ends of the arbors fine pivots are turned. These run either in pivot holes drilled in the plates or bars, or else in jewel holes, to diminish friction and reduce wear. A jewel hole is a small circular plate of garnet, sapphire, or ruby let into the brass of the watch frame. It is perforated in its centre with a fine, true, and polished hole, in which the pivot runs. Fig. 5 shows a wheel and pinion running in jewel holes. A is the wheel, B the pinion, C C the jewel holes.

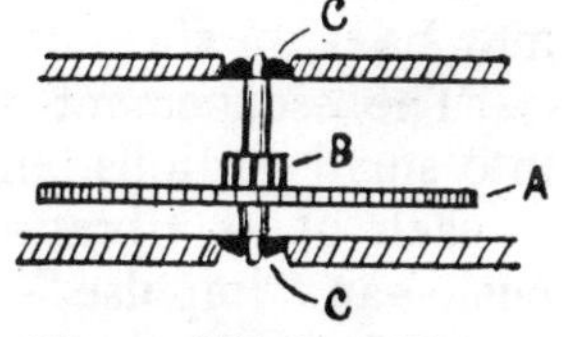

FIG. 5.—Wheel and Pinion.

The teeth of the wheels and the leaves of the pinions are cut to very exact curves, so as to ensure a smooth and even motion when they are running together.

The Escapement and Balance.—It is obvious that, given a mainspring and a train of wheels such as that just described, if the mainspring were wound up the train would run at full speed, the spring unwinding itself in a few moments. Some arrangement is therefore necessary to check it. In a watch this checking mechanism is termed an "escapement," the duty of the escapement being to allow only one tooth of the scape wheel to pass at a time, and that at perfectly regular intervals.

The duty of measuring and regulating the intervals is performed by the balance and hairspring. The balance is a fly-wheel, mounted upon an axis having extremely fine pivots running in jewel holes. It is controlled by a hairspring. The hairspring is a flat spiral of thin steel wire. Its inner end is affixed to a collet upon the axis of the balance. Its outer end is fixed to a stud rigidly fastened to some part of the watch frame. If a balance, mounted and fitted in this way, be given a turn round in one direction and then let go, it will return under the influence of the hairspring and go nearly as far again in the reverse direction until its force is spent. The spring, then, causes it to return again, and it will be kept vibrating for some time before it finally comes to rest. It, in fact, acts in much the same way as a pendulum, which, when set swinging, continues to swing, traversing a smaller and smaller arc each time, until it is brought to rest by friction at its point of suspension and the resistance of the air.

The short intervals of time (generally one-fifth of a second) thus measured by the balance and its spring are always very nearly equal, and, under some conditions, exactly equal, whatever the distance traversed by the balance may be.

The escapement divides up the power of the mainspring into small portions, and delivers these portions to the balance at each of its vibrations, giving it, as it were, a little helping push—an "impulse"—as it comes round each time.

Thus, through the medium of the escapement, the mainspring keeps the balance vibrating, and the balance regulates the running of the train.

The balance and hairspring are well seen in Fig. 1.

The Motion Work.—All the mechanism before described would be of little practical use unless the revolutions of the various wheels could be recorded in some way. The centre arbor, it has been seen, revolves once in one hour. Therefore a hand affixed to it will travel round the dial and serve to show the minutes. By gearing down from this arbor with a pair of wheels and pinions whose combined ratios are as 1 to 12, one pair being usually 1 to 4 and the other pair 1 to 3, and placing another hand upon the arbor of the last wheel, the hours 1 to 12 can be also shown in the usual way. These reducing wheels are termed the "motion work," and are not visible in Fig. 1, being hidden between the dial and the watch plate.

This is a brief description of a simple form of watch movement, and it will easily be believed that a great many different materials are used in its construction and in the processes of repairing it; also that it takes but a very little to upset its action or stop it altogether. Merely to enumerate all the possible faults and inaccuracies to which a watch is subject would fill many pages, and it is the purpose of this book to describe them all in detail and the way to overcome them.

CHAPTER II.

THE MATERIALS USED IN THE CONSTRUCTION AND REPAIR OF WATCHES.

STEEL.

STEEL is used for all the hard-wearing parts of watches, and for parts where strength and rigidity are required—winding squares, barrel arbors, pinions, balance staffs, levers, keyless winding wheels, clicks, and for all screws. It is also used for mainsprings, hairsprings, clicksprings, and innumerable other small springs found in the various parts of keyless and other watches.

The steel used in watches must be of a high quality, and watchmakers generally obtain it in 12-inch lengths from the firms who especially lay themselves out to supply their wants of all kinds. This steel is obtained in the form of *wire*, *rod*, *flat strips*, or *sheet*. The wire and rod can be had in any size, from that required for the smallest balance staff to about $\frac{1}{4}$ in. in diameter, from which a large barrel arbor can be made. The flat strips and square rods are used for making levers, keyless work, small springs, etc. Steel drawn in lengths of a special section can also be obtained for making particular parts, such as "hooking-in" steel for making blocks for hooking mainsprings, "click-steel" for making fusee winding clicks, "pinion-wire" for making pinions, "lever-steel" for making levers, etc. A specially hard steel, "silver-steel," can also be obtained. It is used by some watchmakers to make turning cutters, drills, etc.

Softening Steel.—Steel as bought is "soft," but, being drawn or rolled, is not so soft as it *can* be made, or as it is desirable to make it for some purposes. In the making of a screw, for instance, it is desirable that the steel be first made as soft as possible, so as to take a good "thread," and not injure the screw plate. To soften it, it is heated to a full red and allowed to cool slowly; or it can be made a little softer still by heating to a red, allowing it to cool slowly until, when held in the shade, the red has completely disappeared, and then dipping it in water. This process is called annealing.

Hardening Steel.—In this soft condition it is shaped up by filing, turning, bending, etc., roughly to the required form and size. Before it can be used as a going part of a watch it must be hardened. This is done by heating to a full red and plunging into cold water or oil. Many special mixtures for hardening steel are used by some; also plunging into lead, mercury, etc., is sometimes advocated when great hardness is desired; but there does not appear to be any necessity for the use of anything else but water or oil. When steel is heated to a dull red and hardened in oil, it is tougher, though not so glass-hard as when made hotter and plunged into cold water.

Each method has its uses. For a spring, where toughness and elasticity are required, oil-hardening is best; while for a pinion or for a cutting tool, where extreme hardness of surface is wanted to resist wear and tear, water-hardening should be practised.

Tempering Steel.—Steel just hardened, even by the oil method, is far too brittle for most purposes, and it requires tempering. The application of heat tempers it, that is, it reduces its hardness somewhat, but gives less brittleness and more elasticity. This is only true up to a certain point, at about 570° Fahr. it reaches its maximum elasticity; when heated beyond this point, it rapidly loses both hardness and elasticity, until at about 1200° (dull red heat) it becomes quite soft again.

As bright steel is gradually heated the surface becomes first of a pale straw colour; this deepens into a pale brown, turns to a red, purple, dark blue, pale blue, and then white again. These colours are extremely useful, as showing the temperature, and consequently the temper, to which the steel has been drawn.

Thus, to temper a piece of steel, it is first brightened, to show the colours well, then heated gradually over a spirit lamp flame or upon a slip of brass held in the flame. Blue, that is, the first dark blue that succeeds the red, is the temper to which all springs, screws, pinions, keyless wheels, and other working parts of watches are brought to. The moment this colour appears the steel is removed from the flame, or the blueing slip, and allowed to cool. If a pinion or other part is required to be specially hard, it may be tempered to a red or a purple only.

A drill or a cutter for use on brass or soft steel only is tempered to a pale straw colour; but those for use on tempered steel are made as hard as possible, and not tempered at all.

Good steel may be many times rehardened and softened again without spoiling it, provided it is never overheated. A full red is quite hot enough for all purposes, and, small parts especially, should not be allowed to remain red long.

When hardening extremely fine pieces of steel, such as the drills used for making fine pivot holes, water or oil for dipping need not be used. Indeed, it would be quite impossible to use them, for the steel would be cold before it could touch the liquid. In such a case air-hardening is employed. The drill is heated in the flame, and, when red, withdrawn with a sudden jerk, and the air cools and hardens it. This is termed "flirting." Similarly, it is nearly impossible to soften such a drill or any other piece of fine steel. It could not be cooled slowly enough to prevent rehardening, mere removal from the flame being quite sufficient to harden it again. If such a piece is required to be softened, it must be laid upon a slip of brass upon which is also laid a small piece of brightened steel to indicate the colour. It is then heated, while the piece of "index" steel passes through the series of colours from straw to blue and on to white again. It will then be soft enough to be operated upon or bent as desired.

BRASS.

Brass is an alloy of copper and zinc, a common proportion for good quality brass being two of copper to one of zinc. Common cheap brass has less copper, and is not so strong or so hard. It is distinguished by its very white appearance when broken or filed.

Brass is used very largely in watches. The plates and framework, the wheels, and many other parts being in most watches made of it. Like steel, brass is capable of being softened, hardened, and, to some extent, tempered. Brass rods of all diameters and brass sheet can be obtained ready for the watchmaker. Brass rods, being drawn, are rather hard, but for ordinary purposes of filing and turning, etc., this condition is to be preferred, as it works cleaner than when quite soft. But if it is required to bend or hammer the brass out, it should be softened first.

Softening Brass.—This is done by heating it nearly to a dull red, and allowing it to cool, or, if time is an object, it may be cooled by plunging in water. The result is the same in either case.

Hardening Brass.—Brass is hardened by compression. Any process that tends to compress its particles against one another will harden it. Hammer hardening consists of hammering it with a smooth-faced hammer all over equally until it is sufficiently hard. Unskilful hammering, if long continued, is very liable to crack the metal. Drawing brass wire through a draw plate, decreasing its diameter, will harden it. Twisting a rod also has the same effect. Rolling, as in reducing the thickness of a plate or rod, will harden it. It must not be supposed that any hardening process applied to brass will render it as hard as steel, even in its soft state. But brass can be stiffened very much, and made so elastic that very serviceable springs may be made of it. It is often used for some of the smaller springs in watches.

Very thin brass can be conveniently hardened by simply burnishing it with a polished steel burnisher.

Tempering Brass.—If hammered too brittle, brass can be tempered and made of a more even hardness throughout by warming it, as in tempering steel; but the heat must not be nearly so great. Brass, heated to the blue heat of steel, is almost soft again.

Gold.

Gold is a useful metal in watches, for small special parts. Hard, polished and burnished gold, when working in connection with hardened steel or jewels, causes very little friction indeed. It seems to take a very slippery surface, which is useful in certain cases. Its incorrodibility is also a recommendation. It has been used for hairsprings, but for this purpose its weight is against it. For the screws used to weight compensation balances, for small contact pieces such as curb pins, etc., it is invaluable. Its weight makes it useful for balances, and its extreme ductility is useful in the collets of small wheels, etc., where a good rivet can be formed upon it without undue pressure. As a protective covering for brass, in the form of electro-gilding, it is used in nearly all watches. Also for dial plates and for hands it is much used.

Gold is hardened, tempered, and softened in the same way as brass, but is capable of much more in the way of hammering than that metal, being more ductile.

Silver.

Silver is not nearly such a useful metal for watchwork as gold. Not being easily corroded, it is used for small engraved index plates for regulators, and sometimes for dial plates.

It is worked in the same way as gold. Both gold and silver, being very ductile, can be readily worked up for use in watches. Almost any scrap of metal, such as a jump ring, or a broken brooch pin, can be softened, straightened out, hammered flat, or drawn into convenient sized wire with ease, and serves to make such small articles as screws, pins, or small springs as are required in watches of these metals.

German Silver.

This metal is really a kind of white brass. It is an alloy of copper, nickel, and zinc, in which copper is the predominating metal. It is used in watchwork mainly for the frames and plates of so-called "nickel" movements, of which so many are made in America and in Switzerland. It is worked exactly as brass, but is a little harder than that metal. It is white, and takes a good polish. Movements made of it are not gilt to preserve their surface, but are generally ornamented by "spotting" or etching patterns upon them. They do not readily tarnish.

Platinum.

This metal is only used in watches for balance screws. Here its weight is sometimes an advantage, and often enables a watchmaker to overcome difficulties in timing a watch. It is a very soft metal, and screws made of it are easily damaged by handling.

Palladium.

The only use to which this metal is put in watches is for balances and balance springs for non-magnetic watches, where it takes the place of steel. It is not used pure, but is alloyed with other metals to harden it. Even so, it is but an indifferent substitute for steel, balances made of it being soft and easily put out of truth, and balance springs made of it are heavy and soft, easily distorted and damaged, and unsteady in pocket wear. Still, as it answers the purpose of making the watch incapable of being magnetized in its most vital parts, it is used in spite of its softness. In marine chronometers it is often used for the balance springs, and, it is claimed, has advantages in enabling a more perfect adjustment to be obtained. Here, on account of the thickness of the spring wire, its softness does not matter so much.

Aluminium Bronze.

This is an alloy of aluminium and copper, in which the copper largely predominates. It is hard, takes a high polish, and has a beautiful golden colour. It is a hard-wearing alloy, and is very light. It is largely used for the train wheels of watches, more especially Swiss and American, and for levers and scape wheels, where its wear-resisting qualities and its lightness render it extremely useful. It can be worked much as brass as regards hardening, etc.

Invar.

Invar is a nickel-steel alloy having the peculiar property of hardly expanding or contracting at all with changes of temperature. For this reason it is used for clock pendulum rods and in watches has been used in the construction of compensation balances. It resembles steel in colour and hardness but is rather tougher, and difficult to drill, file, or turn, on that account.

Red-Stuff.

Rouge, red-stuff, and crocus are all one and the same thing, but differ in fineness in the order named, rouge being the finest and the slowest in cutting. They are metallic oxides, and are fine red powders used for polishing metals. The most generally useful in watchwork is the grade known as fine red-stuff, in lumps that are easily pounded up into a paste with oil upon a polishing stake. This should be a dark purple-red in colour.

Diamantine.

This is a polishing powder much used by some, and for some purposes to be preferred to red-stuff. It is white, and sold in small bottles. It has nothing to do with diamond powder, but is only a fancy name given by its makers.

Diamond Powder,

As its name indicates, is the fine dust produced by crushing diamonds. It is not used in watchwork for polishing metals, but only for the materials so hard that no other polishing powder will cut, such as ruby pallet stones, jewel holes, ruby pins, etc. Used on an iron lap wheel, it is very useful for just trimming the corner off a pallet stone and similar purposes. Though used largely by watch jewellers in the manufacture of jewel holes and ruby pallets, etc., the repairer does not need much of it, except as stated above.

Oilstone Dust.

This is oilstone reduced to powder, and is useful made into a paste with oil, for smoothing steelwork preparatory to polishing. It leaves a fine, even, grey surface. It can be also used for "spotting" and "snailing."

EMERY.

Emery, in the form of emery sticks, is very useful. The usual kind of emery stick used by watchmakers is made by glueing fine emery paper upon wood. These soon cut through, but, being very cheap, this does not much matter. Their flatness and general handiness outweigh any disadvantages they may possess. They are obtainable in various degrees of fineness, from the finest, No. $\frac{4}{0}$, to about a No. 4. The coarser ones will rapidly cut a brass plate and keep it fairly flat. They will also cut steel that is too hard to file. The finest produce a passable substitute for a polish upon both brass and steel, especially when they have worn fairly smooth and got the new "cut" taken off them.

WATER OF AYR STONE.

This is a mottled, slaty-coloured stone, used in a block with water for stoning brasswork smooth for gilding or for polishing. A slate pencil, flat on the end, can also be used for the same purpose, and in a confined space has advantages.

CHALK.

Prepared chalk is used dry on watch brushes for cleaning watches and to keep the brush clean. Billiard chalks are a very convenient form in which to buy and use it.

BENZINE.

For cleaning the old oil and dirt from watches and for cleaning off various polishing pastes from parts, etc., benzine is used. It is kept in a glass pot with a ground-in glass cover to prevent evaporation, the vapour being highly inflammable. It can generally be purchased at a chemist's.

PETROL.

Petrol, as used for motor-cars, can be used instead of benzine for cleaning watches. It evaporates quickly and leaves no stain, being fully equal to the best benzine and far cheaper. Like benzine, it is highly inflammable.

OIL.

The oil used for watches is best bought in small bottles, specially prepared by makers of good repute. Attempts to refine oil for watchwork generally end in failure, and experiments in this direction are too dangerous. That prepared for French clocks is useful for mixing the polishing pastes with, for lubricating drills and such purposes, being a pure and rather thin oil. It is also useful for the heavier parts of watches, such as main-springs and barrel arbors, etc. For train-wheel pivots and escapements, only the best watch oil should be used.

METHYLATED SPIRIT.

This is used for the watchmaker's lamp. It burns with a clean, smokeless flame. Articles heated in a spirit-lamp flame are not made dirty and blacked. Spirit is also used to dissolve shellac. When an article has been cemented with shellac for turning, the shellac is removed by boiling in a little spirit in a spoon.

TURPENTINE.

For lubricating hard steel drills, when drilling glass or tempered steel, turps can be used with advantage; also for lubricating a file used upon glass, or on an enamelled watch dial. It cuts more quickly, and lessens the chance of chipping the edges.

ACIDS.

Of the acids, probably the only one useful in watchwork is hydrochloric acid or spirits of salt. It does not act quickly upon brass, but dissolves steel. It is used principally to remove blue oxide from steel by dipping. It is also useful in making soldering fluid. Needless to say, it should be kept far away from the watch board.

MERCURY.

This metal is mentioned, not that it has any use in watches, but that it is a metal that often finds its way into watchmakers' workshops in the form of beads from a broken barometer or thermometer. When it does so, it is extremely difficult to get rid of. It will find its way on to gilt watch plates in an astonishing way, and spoil them by amalgamating with the gilding and spreading all over them. Therefore keep mercury far away from the workshop. It is true it was used in some marine chronometer compensation balances by Loseby, but they are not likely to get into the hands of watch repairers. It may therefore be said to have no use in watchwork. Mercury can only be removed from a watch plate or case by heating over a spirit lamp and evaporating it. The parts then will require polishing again or gilding, as the case may be.

PEGWOOD.

Sticks of dogwood, known as "pegwood," are used for cleaning out the pivot holes of watches and for various small operations. They are bought in bundles ready for use.

PITH.

Elder pith, well dried, is useful for cleaning pivots and other parts. It can be obtained by peeling the sticks themselves and drying, or it may be bought in bundles ready for use.

TISSUE PAPER.

The tissue paper used for holding parts of watches while cleaning, and for wrapping up movements, etc., should be sun-bleached. Such paper is known as silver-paper, and does not tarnish metals like the common sulphur-bleached paper.

SHELLAC.

For cementing pallet stones, ruby pins, etc., in their places, shellac is used. Shellac is a gum, which, when cold, is brittle and hard. It can be melted by gentle heat, at about the heat of boiling water; or it can be dissolved in spirit of wine to remove it. It is conveniently purchased in flakes.

All the materials mentioned in this chapter, and all tools and watch parts mentioned in this book, can be bought from watch material dealers, of whom there are many in Clerkenwell, Coventry, Birmingham, and other large centres.

Watchmakers of to-day find it hard to realize the difficulties under which their predecessors worked. To go no further back than a century, the assortment of tools which they could purchase was very limited. Watch materials were more limited still. They had to make nearly all their small tools, and if a new part were required in a watch, they had to make it outright from the rough brass or steel. Now we buy drills in sets numbered in tenths of a millimetre, broaches, files, and punches in assorted sizes and of every imaginable form. If we require a wheel, a pinion, or a cylinder, we have only to write for it, and obtain it for a few pence with all the difficult work done.

But if the modern watchmaker is spared much of this work, it is well to remember that the great variety of watch movements and the quantity of cheap rubbish now met with, possessing every fault that such things can have, renders the task of repairing watches and keeping them in good order as difficult, and perhaps more difficult, than ever. The introduction of machinery into watchmaking has resulted in the production of watches having some parts that, while turned out by a special machine for less than one penny, will take a good workman the best part of a day to make by hand. In such cases it is well indeed that duplicate parts can be bought, or these watches would not be worth repairing at all.

CHAPTER III.

WORKSHOP, TOOLS, ETC.

Heating the Workshop.—A watchmaker's workshop should be dry and warm. A damp workshop rapidly ruins tools, materials, and watches themselves. A cold workshop nearly always causes poor work and waste of time—poor work because cold fingers cannot do fine work properly, and waste of time because frequent and long pauses have to be made to re-establish the circulation sufficiently to restore the delicate touch necessary to the fingers of a watchmaker. Hot-water pipes against the wall under the work-board appear to be the most efficient means of heating a workshop. They keep it dry, and also keep the workman's feet warm and his head cool. An open fire, generally placed at the opposite end of the workshop to the board, is the worst method of heating.

Light.—A good light is a necessity. Nothing, of course, equals daylight; but, unfortunately, many workshops are so placed that very little good daylight can enter them. Such daylight as can be got is best utilized by placing the work-board right across the window itself. If the sun shines upon the window, ground glass in the lower panes, to keep it off the work and the workman's eyes, is very effective. The most convenient artificial light in most cases is gas. The incandescent gas mantle is hardly suitable, as it has to be moved about continually as the work requires; but this would be a very good light, being quite free from flickering. An ordinary gas jet is best provided with a green cardboard shade, to keep the heat from the workman's head and the glare from his eyes, as well as to throw down the light on to the work. Working with a bare flame is not wise from any point of view. Some workmen prefer working with an oil lamp and shade.

This is cool, steady, and portable, but is a little troublesome, and there is a small element of danger. There are several methods of using ordinary gas-jets; a double-jointed bracket may be used, springing from the back of the board about 8 in. high, or a small pedestal light, as in Fig. 6, is very convenient. This is furnished with a shade, and supplied with gas by means of a length of rubber pipe, which may come from a wall bracket or an overhead fitting. During daylight this pedestal light can be removed, and be altogether out of the way. But, of all lights, the electric is the best, and those in favoured localities where the electric light is available will use no other.

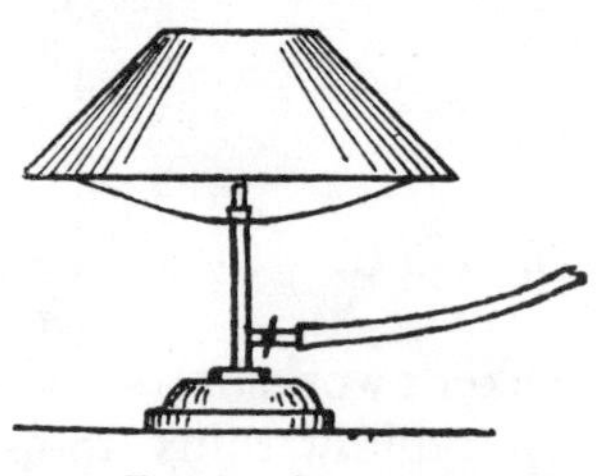

FIG. 6.—Gas Lamp.

The Work-board.—This should be large and roomy. A good width is 2 ft. from back to front, and from end to end for one workman, 3 ft. 6 in. If a watch lathe is kept permanently mounted upon the board, an extra foot of length should be allowed. Inch mahogany is the best for the board, but cheaper woods are often used, even ordinary deal. The latter is soft, and becomes ribbed where the grain runs, causing great inconvenience. On the whole it is far best, in fitting a workshop, to put in a mahogany board.

It is very convenient to be able to either sit or stand to work, and to allow of this the board should be about 4 ft. high. Drawers fitted underneath it are useful for tools, materials, etc.

Tools.—A vice with about 2-in. jaws is a necessity. The old-fashioned hinged vices have quite gone out for watchmakers' use, and some form of parallel vice is used instead. Boley, the maker of so many watchmakers' tools and appliances, makes several patterns of parallel vices, in which the jaws are of hardened steel, and new pairs can be purchased and screwed into position when the old ones become damaged by wear.

Watchmakers use many small tools common to other trades, and also many that are found only in their workshops. Among them the following may be just briefly mentioned.

Pliers, several pairs from about 4 in. long down to $2\frac{1}{2}$ in. (Fig. 7). These, like nearly all watch tools, are far best nickel-

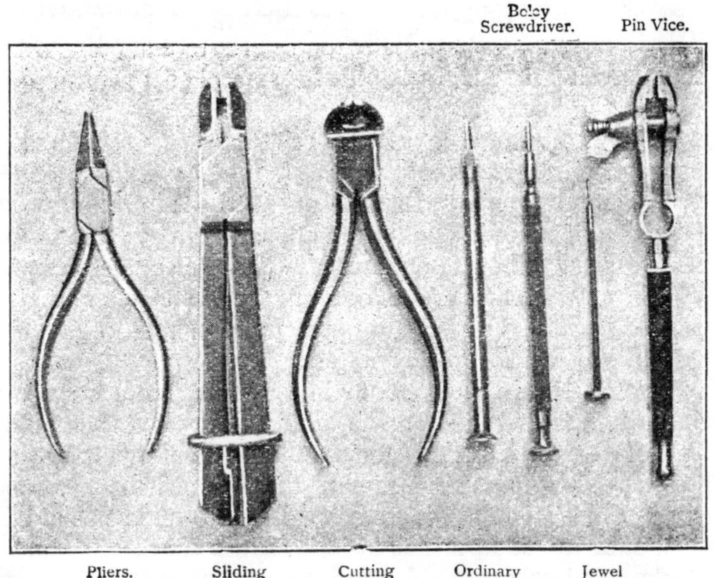

FIGS. 7, 10, 11, 12, 13, 14, 19.—Small Tools.

plated. They are smooth and clean to handle, do not rust, and look well. The smallest pair generally have to be made by filing up the jaws of a larger pair. If the smallest pair is purchased that can be got, the jaws can be heated to a blue in a spirit lamp, and can then be filed up, as shown in Fig. 8, to about half thickness and width. These very small pliers are extremely useful, being something between tweezers and ordinary pliers in size and strength. Of course they are only for occasional use. Another pair of $3\frac{1}{2}$ in. should be converted into brass-nosed pliers, by softening the jaws as above described and filing out recesses on their inner faces. In the recesses pieces of sheet brass or, better still, German silver are soft

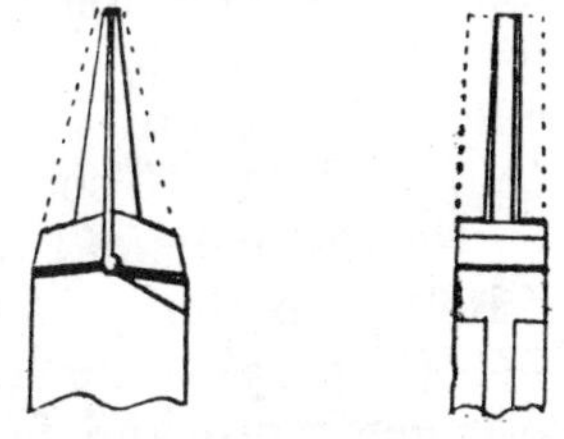

FIG. 8.—Making Small Pliers.

soldered. Some workmen rivet then as well for safety. Fig. 9 shows such a pair, AA being the brass faces. These inner faces should be filed up square and true to close nicely, and their edges should be flush with each other. Brass-lined tools are to handle parts of watches which would be scratched or bruised by the use of steel.

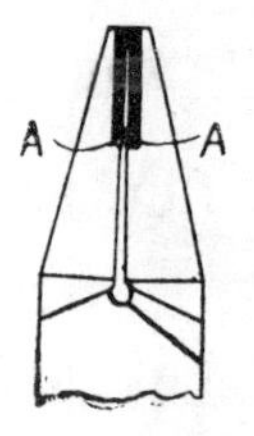

FIG. 9.—Brass-nosed Pliers.

Sliding tongs (Fig. 10) are very useful to hold small parts or pieces of metal while operating upon them with drill or file. An ordinary steel pair should be provided, and also a brass-lined pair, either bought ready made or made from old steel ones worn smooth, in the same manner as the brass-lined pliers. For cutting pins, etc., a pair of 4-in. cutting nippers, like Fig. 11, will be wanted.

Of screwdrivers there should be several. For occasional use on a large screw, an ebony-handled screwdriver with a blade $\frac{1}{4}$ in. wide is useful. The blade should be hardened steel, and kept filed or ground up straight, thin, and true. Properly speaking, this is not a watch screwdriver, but is useful to a watchmaker to use upon his lathe or other tools, or for marine chronometer work etc. A good pattern of watch screwdriver proper is shown in Fig. 12. This is ribbed to give holding power, has a loose top to rest in the hand or on the finger tip, and takes interchangeable blades. These blades slip in a cross slot, and can be removed in a moment. One such screwdriver is useful to take the place of several that are only occasionally wanted. Such, for instance, are an extra wide and thin blade for moving chronometer timing screws, and several split blades for turning nuts, etc. A smaller screwdriver (Fig. 13) of the "Boley" pattern is useful for ordinary small watches, and finally a "jewel" screwdriver for jewel screws (Fig. 14).

Several pairs of tweezers are requisite. For general use the "Boley" pattern, nickel-plated, hollow tweezers answer very well (Fig. 15)—one strong pair and one fine pair. As the fine pair gets worn and filed up a few times, the points get coarse and thick, and they are relegated to "common" work, their place being taken by a new fine pair, and so on. Other useful tweezers are an old-fashioned solid pair, the points of which should be filed off flat, leaving squared flat ends about $\frac{1}{16}$ in. wide (Fig. 16). These form a link between tweezers and the smallest pliers. A moderately coarse pair of *brass* tweezers should also be provided for handling polished steel parts of new watches. Other pairs of specially shaped tweezers will be described in connection with the jobs for which they are used.

Two hammers, one ordinary 1 oz. and one a little lighter, are wanted, both flat-faced.

The large files used should include a rough "potance" file, a rough and a smooth "pillar" file, and a fine file one side

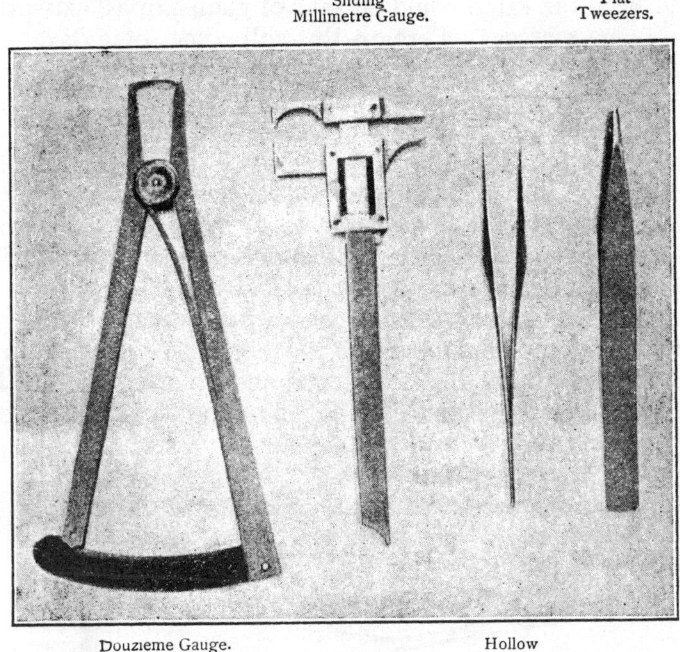

FIGS. 21, 22, 15, 16.—Small Tools.

and burnisher the other of the "potance" size. Of smaller files there are dozens of sizes and varieties, such as crossing files, square, rat-tail, slitting, etc., etc., descriptions of which will be deferred until their use is pointed out.

Drills are generally purchased nowadays from the tool shop, in boxes numbered in series, at a cost of about ½*d.* each. They run from the smallest possible to make up to as large as are ever required in watchwork, and all fit a standard drill stock.

It is therefore hardly worth while to make them. Perhaps the greatest convenience of these drills is that their shanks are all made to gauge. There are three sizes of drill shanks in use. The smallest is for very fine drills, the medium for slightly larger drills, and the large for "clock" drills. To fit them ready-made drill stocks can be purchased, to work with a bow. The various makes of watch lathes also are provided with three drill holders which take them. It is a matter for thankfulness that in drills, at least, we have uniformity. How much better it would be if the same could be said of mainsprings, hairsprings, screws, and glasses! Perhaps this will come some day.

A good set of punches is indispensable. They may be made by the workman, and always used to be until quite lately; but, like drills, they can now be purchased in such very convenient sets that it is hardly worth while making them. Still, there are a few that makers do not yet seem to have included in these sets, and these still have to be made as of old. The latest "staking tool," and set of punches to fit, is one of the most useful tools a watchmaker can possess.

Stakes are nearly always used with punches. They are metal blocks on which to place the work to be operated upon. Some are simply flat steel, hardened and polished; others have graduated series of holes, etc. Fig. 17 shows a graduated stake

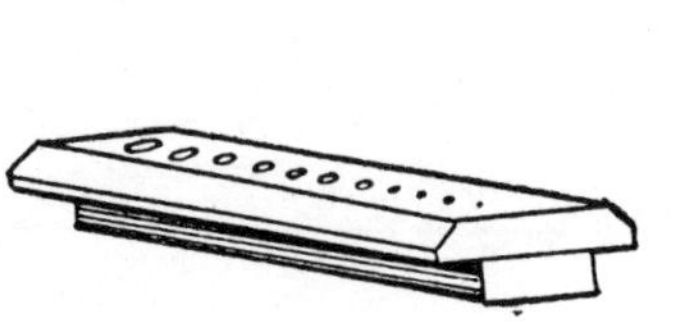

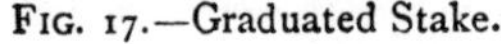

FIG. 17.—Graduated Stake.

FIG. 18.—"Hollow Stake."

shaped to be held in the bench vice. Fig. 18 shows a hollow stake. Others will be described later on.

To enlarge holes, broaches are used. Broaches are tapered five-sided rimers, with five cutting edges. They are obtainable in sets from "pivot broaches" up to ¼ in. diameter. To smooth holes inside, when enlarged to the proper size, round broaches are used. These are merely round, tapered, polished steel arbors, set in handles and graduated like cutting broaches.

For screw-thread cutting, screw plates and taps are required. The great diversity in screws used in different kinds of watches causes endless trouble to the watchmaker. Several screw plates will be necessary, and sets of taps to match them. The "Progress" screws and plates will match most Swiss work and are convenient, inasmuch as screws can be bought finished ready to go into watches, and it is a standard that shows signs of more general adoption.

Pin vices, broach holders, and pin holders are convenient for holding pins while filing them and other similar purposes. Fig. 19 shows a regulation pin vice. The pattern with a hole right through is handy for taking a long length of wire.

FIG. 20.—Pinion Gauge.

Several gauges are necessary for general work, and many special ones for measuring glasses, mainsprings, cylinders, etc., which need not be described here. The general measuring gauges for ordinary filing and turning are three. A "pinion gauge" (Fig. 20), which explains itself. A "douzieme gauge," the name being a Swiss measure = to $\frac{1}{12}$th of a line or $\frac{1}{144}$th of an inch. This gauge, shown in Fig. 21, is useful for small measurements, which can be read on the scale and recorded. For larger measurements an ordinary "Boley" sliding gauge, with vernier, is useful. The size reading to 100 mm. and vernier reading to tenths of 1 mm. is most useful (Fig. 22).

For watchmaking the millimetre gauge is peculiarly suitable. Its unit of measurement is of a convenient size, and for the finest measurements need not run into smaller fractions than tenths, which are easily added, subtracted, or otherwise dealt with. Also the sizing of nearly all modern tools and materials is based on the millimetre. Widths of mainsprings, diameters of glasses, sizes of turning arbors, watch lathe chucks, watch drills, etc.; and in the near future it is possible that *all* watch tools and materials will be sized by it, and so get rid of the utter want of system of the past, under which each maker sized his wares by a series of numbers of his own, having no relation to each other, or to any other series or measure.

Besides the above, a large number of small and special tools are used for particular purposes, many of which have to be made by the workman. These will be described as occasion requires in connection with the jobs for which they are necessary.

Also the appliances for *turning* have not yet been mentioned, being reserved for a future chapter.

Eye-glasses.—The preceding list of tools will be closed with a reference to the *eye-glass*, without which it is useless to attempt watchwork.

It is erroneously supposed that watchwork is very trying to the eyes. No doubt it would be so if the eyes were not assisted by suitable glasses. In fact, it would seem that constantly using the eyes to see small objects, provided the light is always good, acts upon them as exercise does upon the bodily muscles, and they become strengthened. A watchmaker's eyes are probably more capable of seeing fine detail upon everyday objects than those of an ordinary person, for the reason that they are daily exercised and trained. He does not strain his vision. When an object is not easily seen by the unaided eye, an eye-glass of moderate power is used, and it is rendered easy at once. Still finer objects are viewed in more powerful glasses with the same result. It is constantly trying to see objects with insufficient light, and straining to see detail too small to be easily seen by the unaided eye, that injures the sight.

Given sufficient light and magnifying power to enable objects to be easily seen, the eyes cannot be injured. The writer's experience has been that writing and reading, as ordinarily carried on in private rooms and in offices, is far more trying to the eyes than working all day at watchwork. It is also a matter of common observation, that, in strong contrast to the student of books, the young watchmaker's eyes give him no trouble; while in after years he manages without spectacles as long as the generality of other folks.

The ordinary glass used is of about 4 in. focus. This magnifies about 2½ times, and is quite sufficient for the ordinary run of watchwork. But for viewing pivots or jewel holes to detect minute signs of wear, a 2-in. or 1½-in. focus glass is useful. Such a glass magnifies about 6 times.

The ordinary 4-in. glass purchased at the tool shop in a horn or vulcanite mount for 1*s*. is considered sufficient by most workmen. It consists of a single bi-convex lens. Such a lens, every optician knows, has many imperfections. First, it is not achromatic, and fringes all objects *slightly* with prismatic colours, taking from their sharpness and contrast with their background. Second, such a glass has considerable spherical aberration; that is to say, all the light collected by it is not focussed at one point. The marginal rays are of shorter focus than the central rays. To a small extent the latter fault can be corrected by cutting off the extreme marginal rays, the worst offenders, by means of a circular stop of brown paper screwed into the lens mount in contact with the glass. Such a stop may have a central hole of $\frac{3}{4}$-in. It causes a little loss of light, but greater sharpness.

Both chromatic and spherical aberration can be somewhat reduced by using a double lens—two similar lenses in one mount, like Fig. 23. Such lenses are sold, and are miscalled

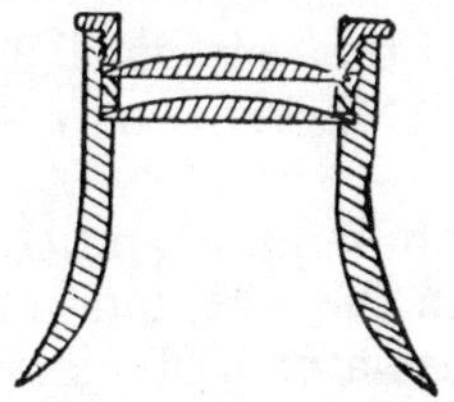

FIG. 23.—Weak Double Eye-glass.

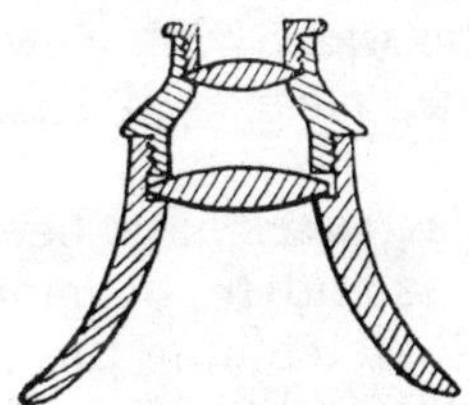

FIG. 24.—Strong Double Eye-glass.

"achromatic" lenses; but they are not much improvement when used for the ordinary 4-in. glass. On the other hand, the more powerful glass of $1\frac{1}{2}$-in focus is greatly improved by being made in this way, the usual pattern being like that shown in Fig. 24.

A 4-in. focus lens should cover a field of view of 2 in. diameter easily. If a watch dial be laid upon the board in a good light and viewed by the aid of a glass like that shown in Fig. 23, it can all be seen at once, but it will not all be equally sharp. If the centre is focussed, the edges will not be quite, and also the black figures on the white ground will be fringed with blue, red, and yellow. The constant use of such a glass must be more tiring than one in which these defects are absent.

To get over the difficulty, the writer has used for many years a really achromatic lens in which the spherical aberration has also been neutralized. Such a glass is shown in

Fig. 25. It consists of a bi-convex lens of crown glass cemented with canada balsam to a plano-concave lens of flint glass. To have this made specially would be expensive, but, fortunately, such lenses can be purchased cheaply in the form of object glasses for small opera-glasses. A pair of these only costs a shilling or two. One of them about $\frac{7}{8}$ in. in diameter and 4-in. focus can be placed in a horn mount from which the common single lens has been removed. To accommodate its extra thickness, the seating will have to be turned out in a lathe or mandrel, with a slide-rest cutter. The lens mount can be chucked by cementing it with sealing-wax on to a slip of flat wood and clipping the wood in the dogs on the face plate.

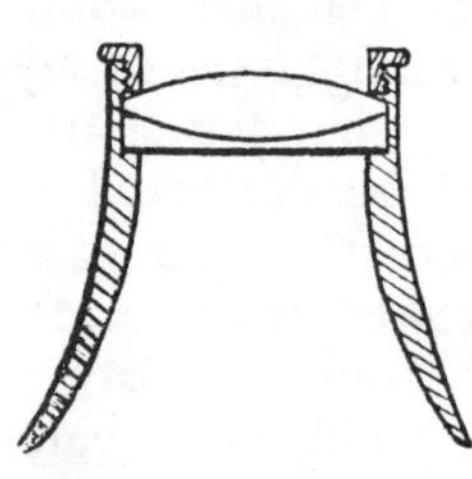

FIG. 25.—Eye-glass with Achromatic Lens.

The watch dial, viewed by the aid of this lens, is all in focus at once, clear and sharp, with a total absence of coloured fringes.

Eye-glasses have been mounted in aluminium and cork for lightness and to be more easily held in the eye, but the best way of overcoming this difficulty is to use a coil of piano wire, as shown in Fig 26. This is passed around the head, holding the mount lightly against the eye. When not required it is pushed up on to the forehead out of the way. A glass held thus is always at hand—no searching for it all over the board—saving much time. It is also open to question whether the habit of holding the glass by contracting the muscles of the eye improves the vision. It is certainly tiring. Sometimes considerable annoyance is caused by the lens steaming. This results from evaporation from the face condensing on the inside surface of the lens, and takes place more especially when the glass is just taken up to use on a warm day. It will be minimized, if not altogether prevented, by making several large ventilation holes in the horn mount, allowing free access to outside air to dry off the condensation upon the lens.

The foregoing is written more for the professional than the amateur. For his work, only a small selection of tools will be needed, and an ordinary eye-glass will answer. But only occasional as may be the jobs which he takes in hand, a good light and an eye-glass are both necessities, and such of the tools enumerated as appear to meet his special requirements may be purchased.

The firms who supply watchmakers with their tools issue very full and well-illustrated catalogues describing each appliance minutely. It has, therefore, not been thought necessary

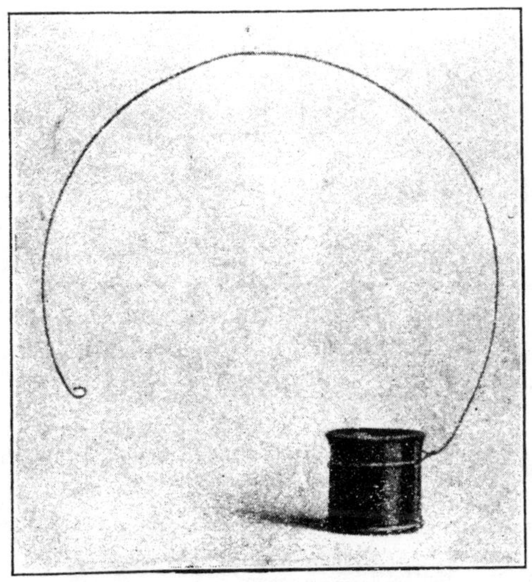

FIG. 26.—Wlred Eye-glass.

to give here more than mere references to them. Such illustrations of tools as are given are more for the purpose of indicating the particular kind of tool referred to than to describe the tool itself.

CHAPTER IV.

THE USE OF TOOLS.

Flat Filing.—Filing is the first operation to be learned by an apprentice. To file flat and true is an art difficult to acquire, but very necessary to a good workman. Novices attempting to file often seem to imagine that a file is to be used as an emery stick, or rubbed backwards and forwards over the work with as much pressure as can be applied, until the metal is removed by abrasion.

This is far from the truth. A file is a cutting tool, resembling very much in action a carpenter's saw. Its teeth are of the same general shape. Consequently the file only cuts one way, and that is on the outward stroke, just like the saw. Suppose a piece of brass of uneven surface to be screwed in the vice. The potance file is taken in the right hand, and the forefinger of the left is placed upon its tip to help to keep an even pressure from end to end of the stroke. The file is first carefully held level, and a forward stroke is made, with just sufficient pressure to cut freely. All the attention should be directed to keeping the file level from the commencement to the end of the stroke. When the end of the stroke is reached, the file may be lifted off the brass and drawn back for another cut, seeing that it is level before commencing again; or it may be drawn back lightly just in contact with the brass and another stroke given. This latter method is the one generally adopted when continuous filing has to be done to reduce a piece of metal, as it has the advantage of not disturbing the level of the file so much at each stroke as lifting the file off the work does. Two or three such strokes with a sharp file should make the uneven surface of the brass flat.

The tendency of a beginner is to round the surface of the work, leaving it high in the centre. This is caused by not regulating the pressure at each end of the file. If too much pressure is applied to the file handle at the commencement of the cut, the handle will be depressed. The secret is to use hardly any downward pressure on the handle at first, but merely to use the handle to push the file forward, exerting all the pressure needed to cut (not a great deal) by means of the finger on the file tip. At about mid stroke the pressure on file handle and file tip must be more nearly equal, and at the end of the stroke all the needful pressure is put on the handle and none on the tip. Thus, the pressure upon the file handle begins at nothing and finishes at a maximum, while the pressure of the finger on the file tip begins at a maximum and finishes at nothing. The art of automatically regulating these pressures to a nicety is the art of filing flat, and it is only to be acquired by much careful practice, taking slow and deliberate strokes.

A good way to practise filing flat and true is to take a 1-in. length of $\frac{1}{4}$ in. brass or steel rod and file it square throughout its length, just leaving the original surface of the round rod visible as a thin line down each corner.

Strictly speaking, a surface is either flat or it is not. There are no degrees of flatness. Judged by this standard, the man who can file flat does not live. By flat filing is meant producing a surface not visibly convex, one that will bear the application of a straightedge without the latter rocking on its centre. When a piece of metal, of the size usually operated upon by a watchmaker, has been thus filed "flat," there are several methods of making it still flatter. The metal may be laid upon a piece of cork screwed in the vice, and filed again with a file of finer cut, allowing the file to rest flat on the metal and never leave perfect contact with it. Or the metal may be laid upon the tip of the forefinger of the left hand and filed as on the cork. If too large for that, it may he held in the finger and thumb against the file. In any of these cases the principle is that the metal operated upon is made to follow the unequal motion of the file and is not rounded by it. Surfaces filed in this way are very nearly flat.

Pin Filing.—Flat filing is not all that is required, as the watchmaker has also to file *round* pins and rods. Round is here used in the same relative sense as "flat" has been, and means passably round, not visibly oval or having flats upon it. For this purpose a pin vice is used. Into this the wire or rod is screwed, taking care that it is central and straight. In the bench vice a block of boxwood is screwed, having several grooves of various depths in it. In one of these grooves of about half its depth the wire is laid, the pin vice being held level between the thumb and forefinger of the left hand. The file is taken in the right, and as the outward or cutting stroke is made the pin vice is revolved between the thumb and finger by a twirling motion in the direction to meet the file, being given at least a complete revolution during the stroke of the file. On the back or non-cutting stroke the file is just pressed sufficiently on the pin to keep it in its groove, and the pin vice is revolved in the contrary direction. The process is repeated until the pin has been evenly tapered down from its full diamater to almost a needle point. If the revolution of the pin vice is stopped during the stroke of the file a flat will be made upon the pin. If the pin vice is not revolved sufficiently the pin will become oval. Fig. 27 shows how the pin should lie in its groove on the boxwood.

FIG. 27.—Position of Pin Vice.

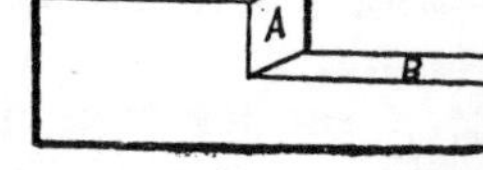

FIG. 28.—Filing a Square Shoulder.

The Use of Files.—Watchmakers' files generally have one if not two "safe edges." A safe edge is plain, having no teeth cut upon it.

When filing a shoulder, as in Fig. 28, if it is desired to cut deeper and not further along the steel, the safe edge is used next the surface A; while if it is deep enough, but wants backing more at A, the file is used up on edge, the safe edge resting on the surface B.

Brass and other soft metals require a sharp new file to cut freely. A worn file will only slip over the surface and burnish the metal. On the other hand, steel, and especially tempered steel, cuts best with a worn file, one in which the extreme sharpness has been taken off the teeth. Therefore a new file should first be used for brass only, until too worn for that purpose, then it is passed on for steel use, and a new one started for brass. A rough potance file, which is only occasionally used to rapidly reduce a large piece of metal, may be chalked on one side to distinguish it, and the chalked side kept for use on brass, the other being used for steel. This insures having one sharp side for use on brass.

Steel that has been hardened and tempered requires care in filing to see that the file really cuts and does not slip. Considerable pressure is required and very slow strokes. By allowing a file to slip upon blue tempered steel, it is quite possible to so burnish and harden its surface that the file cannot be persuaded to start cutting again. Also it ruins the file to let it slip.

Drilling.—Drills are used in two ways by watchmakers. They may be held in a stock with a ferrule attached and rotated by a bow, or be placed in a lathe. The first method revolves the drill alternately in both directions; the second gives a continuous motion in one direction only. Fig. 29 shows the arrangement used in drilling with a bow. A is the drill, fastened by a clamping screw, B, into the stock C. D is the ferrule for the bow to encircle. The centre at the ferrule end works in a centre hole in the vice jaws. The work to be drilled, E, is held up to the drill by hand only. The bow varies according to the size of drill in use. For pivot drills, the smallest, a bow of thin whalebone about 9 in. long, strung with a stout horsehair, is strong enough. A larger whalebone bow, strung with strong thread, is used for medium drills, and for the largest clock drills a cane bow, strung with twine or gut. The bow string is given one turn around the ferrule.

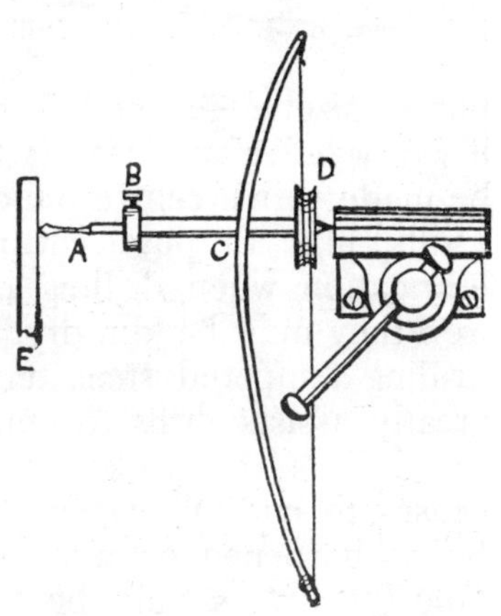

FIG. 29.—Drilling with a Bow.

This method of drilling is handy for odd jobs in which no particular truth is required; but if anything has to be drilled straight and true, the lathe is to be preferred. Drills for use with a bow are generally sharpened on both sides, as in Fig. 30. A cutting edge of this wedge shape does not really cut; it scrapes. It simply has the advantage of cutting, or scraping, equally in both directions. Since lathes have come more into use, drills sharpened to cut only in one direction, as in Fig. 31, are common, and do for both bow and lathe. When used with a bow, these drills cut on the forward stroke and run back, not cutting at all on the back stroke. On the whole, even with a bow, they will be found to cut better than those shown in Fig. 30.

As seen in Fig. 31, the "point" of a drill is not a point,

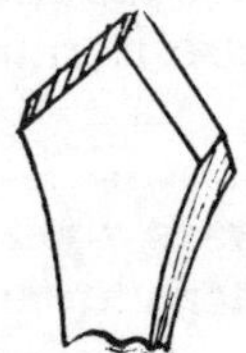

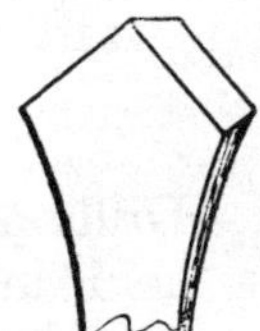

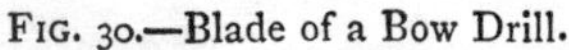

FIG. 30.—Blade of a Bow Drill. FIG. 31.—Blade of a Lathe Drill.

but a chisel edge; and it requires a centre to start in, or else it will wander and not start where required. This centre may be made with a centre punch, or by a pointed chamfering tool. Drills, like all pure cutting tools, require plenty of oil for lubrication when drilling ordinary metals. For drilling cast iron they may be run dry, except in the smallest sizes. For drilling tempered steel, turps is the best lubricant. Pressure greatly assists drills to cut, and as much should be applied as the drill or the work can safely stand. The moment a drill ceases to cut by reason of its edge becoming dulled it should be re-sharpened on a smooth-cut oilstone. In drilling a deep hole the drill should be constantly drawn out and cleared of cuttings, or it may choke and break in the hole. This is especially true of drilling soft metals, such as zinc, copper, soft gold, and silver, etc., also for substances like vulcanite, ivory, jet, etc.

Broaching.—Broaching requires good lubrication and constant clearing of cuttings. A broach should not be forced, but just worked with sufficient pressure to continue cutting. The smaller sizes of broaches are best held in holders or a small pin vice. The larger ones may be driven into wood handles.

Cutting Screw Threads.—Taps for cutting screw threads should not be forced, and require plenty of oil. In using a tap, a great deal depends upon the hole being of the correct size to begin with. A hole a little too small will break the tap if much pressure is applied. A tap should be worked in very gradually. A half turn forward, then eased back to clear cuttings and allow the oil to flow; then forward another half turn, and so on. If a tap squeaks, it is a sure sign it is not cutting, and that if pressed it will break. The same remarks apply to cutting threads with a screw plate. Small screws or pins can be threaded by holding them in a pin vice. The plate is held in one hand and the pin vice in the other, and with care no strain is put upon the thread. Larger work is best fixed upright in the bench vice, while the plate is worked on with both hands to balance the strain as much as possible.

Watchmakers' screw plates, that is, sizes up to $\frac{1}{16}$ in. diameter, generally have no cutting edges; they only burr or force a thread upon the screw. Small screws made thus have harder and smoother threads than when cut by a true cutting plate.

CHAPTER V.

TURNING.

Using the Turns.—Turning with a bow and "turns" will be first described. Fig. 32 shows a pair of Swiss turns. They

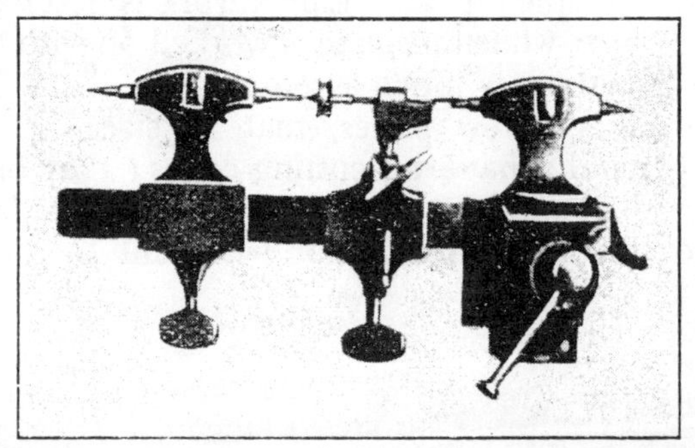

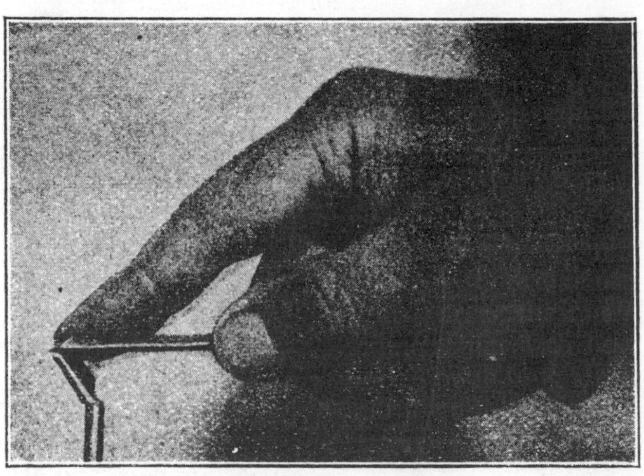

FIG. 32.—Turns; and how to hold the Graver.

are provided with "runners," which are lengths of steel or brass rod, having centres, between which the objects to be turned are placed. The work is rotated by means of a bow, as in drilling, the same bows being used. An adjustable ferrule is attached to the work. Fig. 33 shows such a ferrule. Or the work, if it has a central hole, like a watch hand or a roller, is placed upon a turning arbor like Fig. 34. The various ways of turning

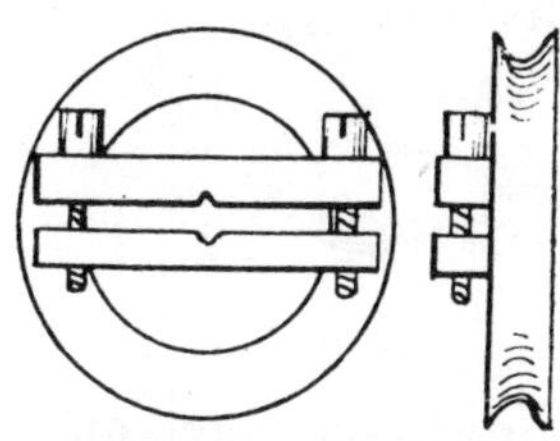

FIG. 33.—Adjustable Ferrule.

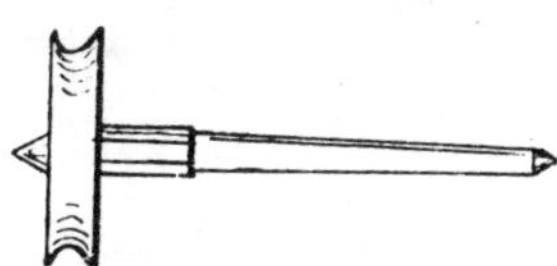

FIG. 34.—Turning Arbor.

work and fitting up the turns for different jobs will be described later on. In this chapter the method of turning only will be considered.

Adjustable ferrules, like that shown in Fig. 33, and of other patterns also, can be bought in sets from very small sizes upwards. Also sets of turning arbors, like Fig. 34, made of hardened steel with brass ferrules attached, are sold in running sizes to take anything from a roller to a clock wheel.

The hand rest of a pair of Swiss turns, as bought, is too wide for most purposes, and is in the way. It will be best to at once cut it off at each end, as shown in Fig. 35 by the dotted lines.

The turning is done by gravers. Gravers are of hardened

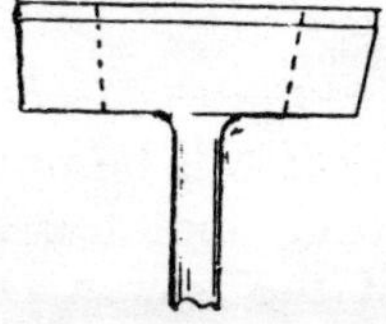

FIG. 35.—Hand Rest.

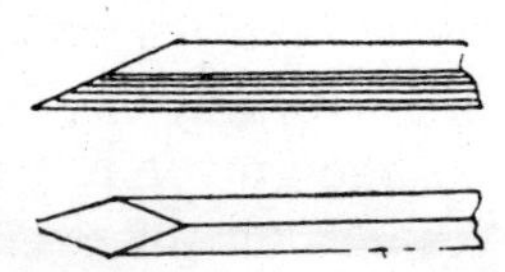

FIG. 36.—Graver Point.

steel, square in section, or diamond-shaped. The diamond-shaped gravers are known as "lozenge" gravers, and are useful where a long narrow point is desired. The graver is sharpened to an angle, as shown in Fig. 36. It will be seen to have two bright sides and two black ones. The black ones should come on the top and the bright ones underneath. A graver is prepared for turning by first grinding its face down to the angle shown in Fig. 36. Then it is rubbed smooth and flat on an oilstone, upon the ground diamond-shaped face. When sharp up to the edges, the two bright undersides are just given two or three rubs on the stone to make the edges quite smooth. Gravers should be held in handles, like Fig. 37.

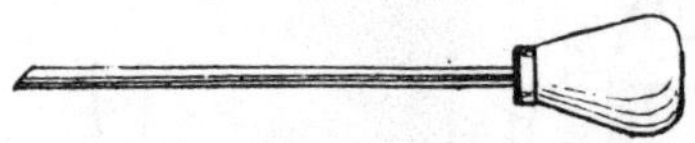

FIG. 37.—Graver in Handle.

Suppose a piece of brass rod $\frac{1}{4}$ in. in diameter and $\frac{1}{4}$ in. long be drilled from end to end and the hole broached out tapered and smooth. This is pushed tight upon an arbor, like Fig. 34, and placed in the turns between the plain centred runners. The medium bow is used, and the turns are held in the bench vice as in Fig. 32, which shows all in position for turning. The bow is held in the left hand and the graver in the right, resting upon the hand rest, as shown in Fig. 32. The rest should be adjusted as close to the work as possible. Fig. 38 is a view from above, showing graver point and work

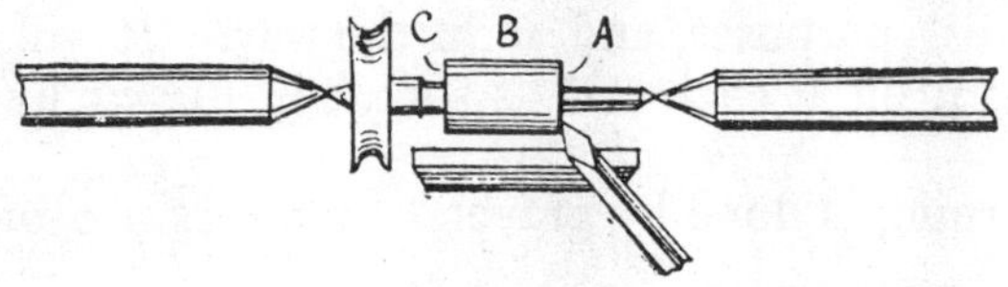

FIG. 38.—Position for Turning.

in position for turning the face A true. It will be noticed that the point of the graver is used for cutting, not the flat edge. Cutting is done on the down stroke of the bow. During the up stroke the graver point is eased off from the work, so as to be just free from it, by an almost imperceptible movement of the hand and wrist. The graver is not slid backwards and forwards upon the rest to free it, but is turned aside an almost immeasurable amount without actually shifting it on the hand rest.

To begin with, the face A will be out of truth and the first few cuts will be uneven and jerky, the graver point only touching the projecting parts; but after a while it will be turned true, when the graver point should take off a clean shaving, like Fig. 39. The face A being turned true and smooth, the face B, Fig. 38, is commenced, holding the graver as in Fig. 40.

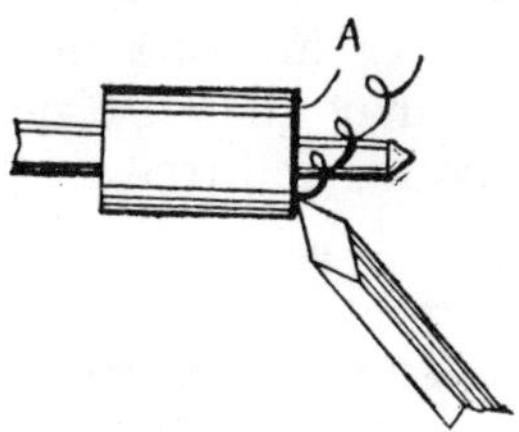

FIG. 39.—Cutting the Face.

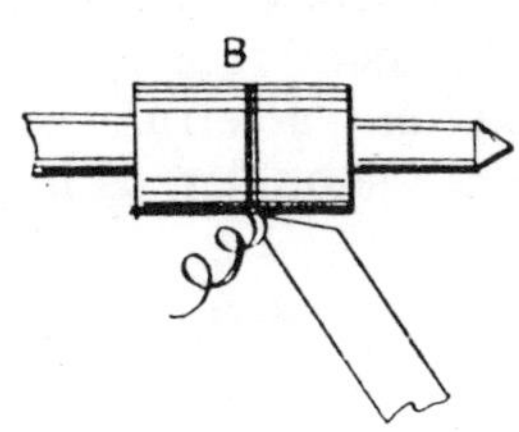

FIG. 40.—Cutting the Circumference.

To turn C, the arbor may be reversed as in Fig. 41, or the brass may be taken off the arbor and reversed. The latter method will not make the work so true as the former, as, the hole being tapered the wrong way to fit the arbor, the brass will not go on exactly as before. In reversing the arbor, as shown in Fig. 41, the bow is still held in the left hand and

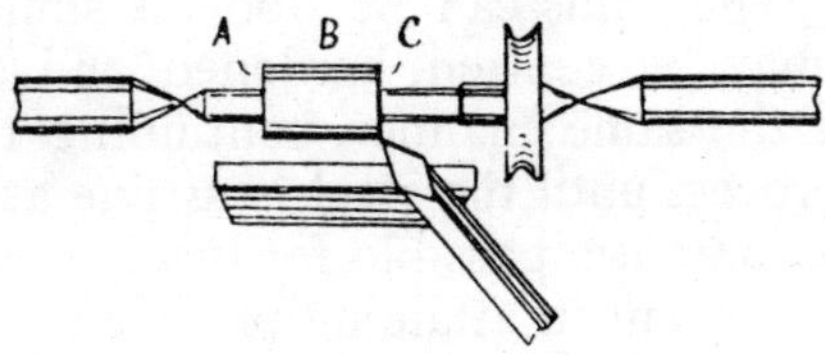

FIG. 41.—Position reversed to cut the Back.

graver in the right. Care should be taken that oil is applied to the centres, or the arbors and turn runners will be spoiled.

Brass requires a sharp graver to cut clean and smooth, and a fairly high speed of revolution. To attain this, use a long bow, and use its entire length at each stroke.

Steel should not be turned so fast, especially tempered steel, or it will glaze and refuse to cut at all.

The graver should be applied level with the centres, that is, the graver must point to the centre of the work. If below, it will tend to spring the work up and roll it upon the graver point. This would be certain to either break the point or spring the work out of truth if it is slender. If above the centre, it will not cut. Fig. 42 shows what is meant. The graver in this figure is shown held as in Fig. 40.

Brass is the best metal to practise upon at first. When able to turn a brass cylinder smooth and true on all faces as just described, a length of soft steel may be taken in hand. A piece of steel rod $\frac{3}{16}$ in. diameter and $1\frac{1}{2}$ in. long may be centred and a ferrule affixed to it. It may then be turned down to a shoulder, as in Fig. 43, A, smooth and true. Then

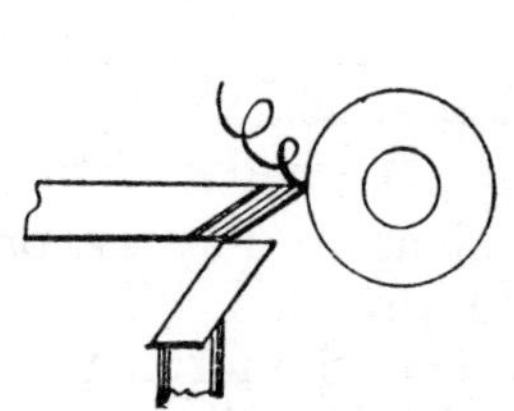

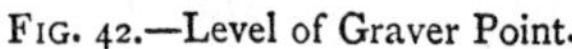

FIG. 42.—Level of Graver Point.

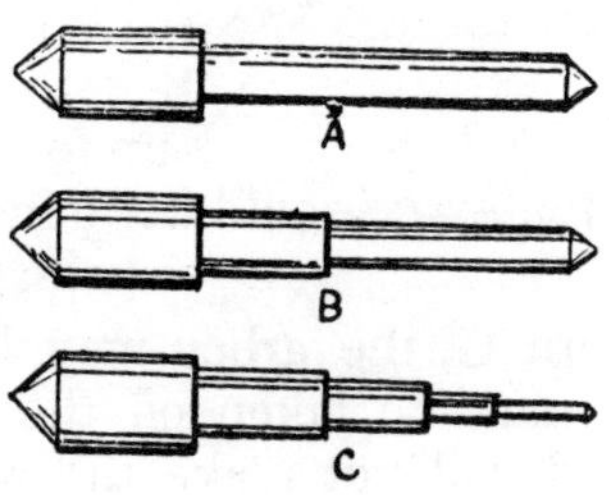

FIG. 43.—Turning Square Shoulders.

a second shoulder may be turned, as B; then a third and a fourth, as C. When this can be done, a similar or a smaller piece of steel may be centred, hardened and tempered blue, and treated in the same manner, continuing the shouldering and reducing process until the steel is as fine as a watch pivot. This is the best practise possible for turning, and many hours may be profitably spent in attaining proficiency in it.

The Watch Lathe.—Until about twenty years ago the turns as here described were used by all watchmakers, but since that time watch lathes have been perfected, and have made quite a revolution in the watch repairing shop.

The great disadvantage of the turns is the alternate motion and consequent break of the "cut" of the graver each time the bow is raised. The continuous motion of the lathe renders turning easier, more rapid, and more certain. Another disadvantage of the turns is that only work can be turned in them that can be placed between centres or upon an arbor. Work that required chucks to hold it, or a face plate, could not be done; hence a mandrel was required as well as turns for such work. A mandrel is a lathe having a fixed face plate for holding work and a slide rest and hand rest.

The watch lathe combines in one tool the turns and the mandrel, as well as having a multitude of chucks to hold nearly all parts of a watch. And it turns all with a continuous motion. For a young watchmaker buying his tools, it will be as cheap, and very much better, to begin by purchasing a lathe and leave the mandrel out. The many chucks with the lathe will also enable him to dispense with the pivot centring tool, the Jacot tool, and the screw-head tool. But the turns themselves cannot be altogether discarded. For many small jobs they save a great deal of time, and really cannot very well be done without. In the workshop of the up-to-date watchmaker the watch lathe and the turns should be found side by side.

There are two principal patterns of watch lathe, the

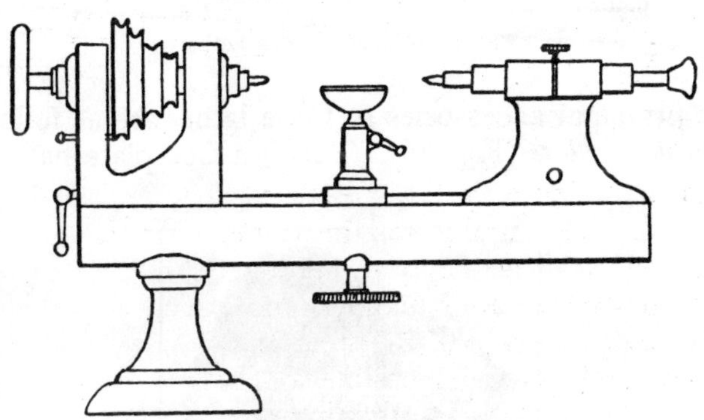

FIG. 44.—American Watch Lathe.

American and the German. The American pattern is the larger of the two. It stands upon a pedestal foot and has a broad solid bed. Its centres are about 2 in. high, taking a 4-in. mandrel face plate, large enough for any watchwork. Where board space is available, such a lathe is the best, and it can always be erected ready for work, and be protected when not in use by a wood box-cover.

The German pattern more resembles an overgrown pair of turns, and is smaller and lighter. It is fixed in a pedestal foot, or screwed in the vice like a pair of turns. Its bed is a bar, and its centres only about $1\frac{5}{8}$ in. high, taking a face plate $3\frac{1}{16}$ in. in diameter. This is large enough for most jobs, but occasionally it is not large enough, and causes trouble. Each pattern is provided with the same kind of chucks, etc., and, except in point of size, there is not much between them.

Fig. 44 shows an American pattern lathe, and Fig. 45 a German one.

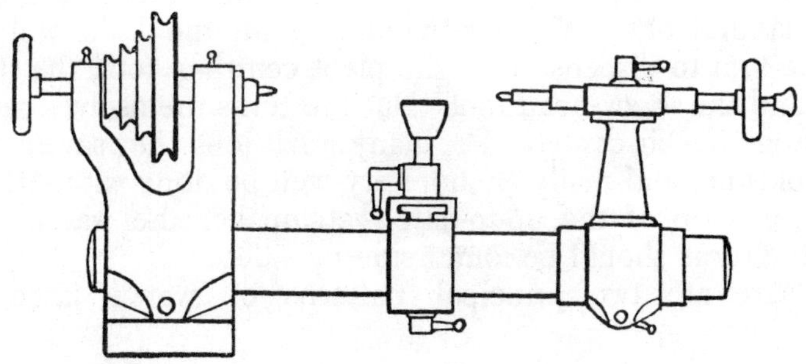

Fig. 45.—German Watch Lathe.

The principal accessories to these lathes are as follows :—

A *mandrel plate* (Fig. 46). This is a face plate having three

Fig. 46.—Mandrel.

slots and three dogs to hold flat plates, etc., for turning on the face. In some lathes it is a separate headstock, as in Fig. 46, to ensure absolute truth; in others it is in the form of a chuck. It is provided with a "pump centre." This is a loose centre runner, by which work upon it is accurately centred from the back.

A *slide rest* (Fig. 47) to use in conjunction with the mandrel

FIG. 47.—Slide Rest.

or any other chuck. This is provided with two slides. The lower cross slide is fixed; the upper longitudinal slide swivels, and can be set to any angle required. A set of cutters to fit it may be bought ready made.

A *wheel-cutting apparatus* is made to suit most lathes. It

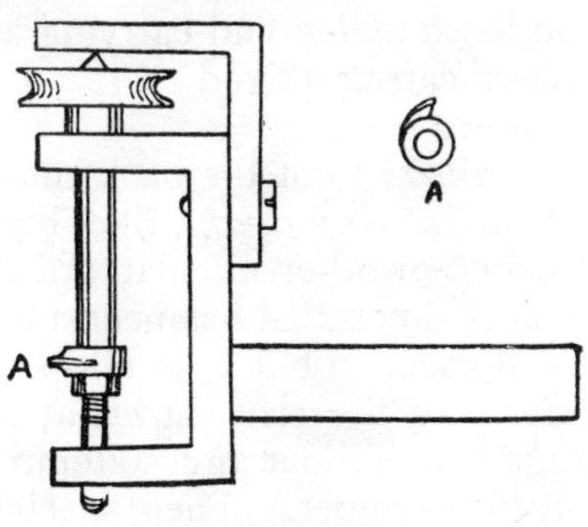

FIG. 48.—Cutter Frame.

is rather elaborate and expensive; but a simple cutter frame that can be fixed in the tool holder of the slide rest, devised by the writer, is shown in Fig. 48. The frame is filled up from a brass casting. The arbor is tempered steel, and put in from the top, being held by the upper cock, which is adjustable and presses on the top centre only. The cutters are shaped as shown at A, and may be single-bladed "fly" cutters, filed up and hardened, shaped up when hard by polishing with oil-stone dust and oil on a steel polisher, to fit a tooth space accurately. The cutter is held by a nut screwed against it, and may be itself also screwed on, the nut being a lock nut. It is driven by a thread band from a 6-in. wood pulley on the distributing pulley axis, to get as much speed as possible. These cutters cannot be driven too fast.

A *hand rest* like that provided with turns, and for the same purpose.

Loose pulley runners or chucks for turning between dead centres, like Fig. 49. The centre upon which the work turns

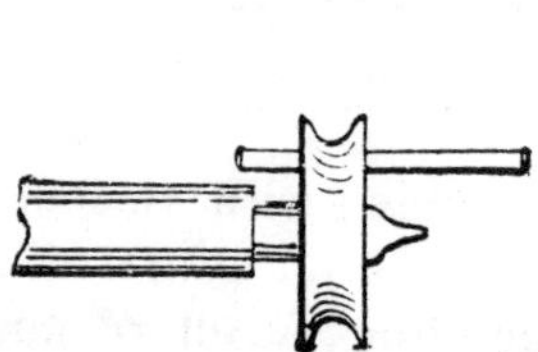

Fig. 49.—Loose Pulley Runner.

Fig. 50.—Carrier.

is stationary. The loose pulley and carrying arm only revolve, driving the work by a carrier affixed to it. Fig. 50 shows one of a set of such carriers.

A set of *turning centres* to fit the back runner.

A set of five *wheel or step chucks*, like Fig. 51, for holding wheels, barrels, covers, or other circular articles. A (Fig. 51) shows a step or wheel-chuck that has been turned down to half size. It is time well spent to buy a second set of these chucks and turn them down to the size shown at A. They can be done with the slide rest without any softening or other treatment, being about blue temper. Their special use is for holding small wheels when the pivots require turning or polishing, etc. If such wheels are placed in the full-sized wheel chucks they are buried in the chuck, and the pivot cannot be seen or got at to turn properly.

A set of *split chucks* (Fig. 52) for holding wire, pinions, arbors, etc. These are sized in tenths of a millimetre, from No. 4 = $\frac{4}{10}$ mm., upwards, a useful assortment being 4 to 40.

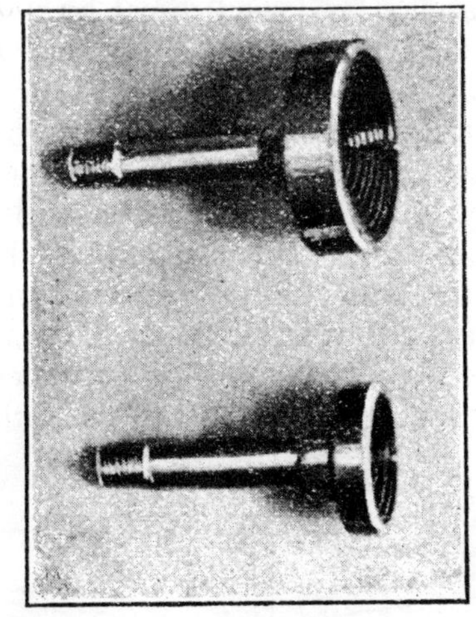

Fig. 51.—Step Chucks.

An *adapter chuck* and shellac chucks to fit it, for attaching parts by shellac which cannot otherwise be held conveniently.

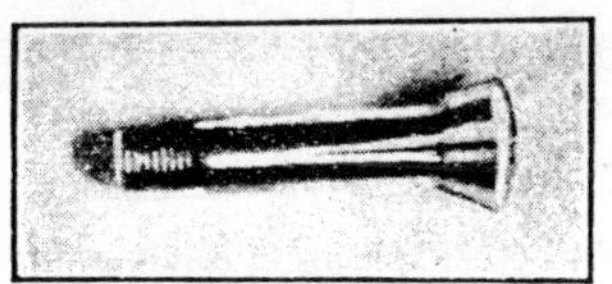

Fig. 52.—Split Chuck.

A *drill holder* to fit the tail stock, with clips for standard sized watch drills.

Besides these, many other chucks and appliances are provided, some of which will be described later on as the use of them becomes apparent.

The chucks are of a peculiar pattern. They do not screw into the mandrel noze like an ordinary lathe, but fit in a plain turned seating, held by a key or feather from turning round.

and drawn in by a drawing-in spindle from the back. This method answers the double purpose of ensuring absolute truth and automatically closing the chucks upon whatever work is placed in them.

A watch lathe may be driven direct by a hand wheel fixed under the board, or by a foot wheel fastened to the floor; or a foot wheel may be inverted and fastened to the underside of the board, the treadle driving by a cord on to the crank. The treadle in this case can rest upon the floor or be hooked upon the cross rail of the stool. The latter is a very nice way of working, leaving the workman seated upon his stool as usual, with both hands free and the poise of his body undisturbed.

Generally, when a foot wheel is used, the lathe is not driven direct, but through a distributing pulley placed at the back of the board. The arrangements in use are outlined in Fig. 53.

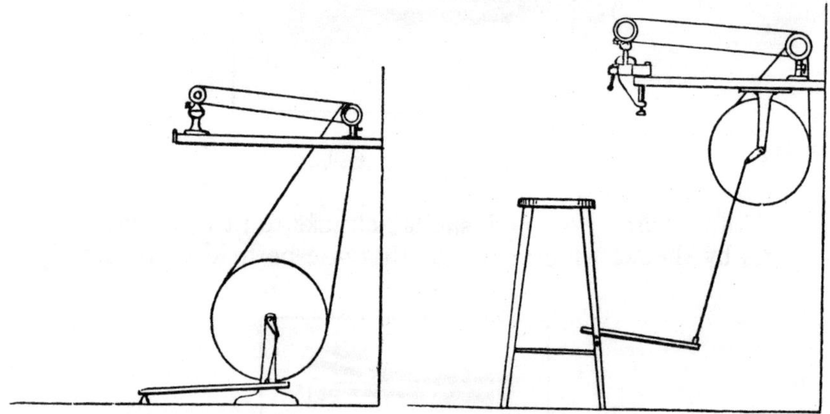

FIG. 53.—Methods of driving Lathes.

For a driving band from foot wheel to distributing pulley, grandfather clock gut does very well. From distributing pulley to lathe, vienna clock gut or cotton cord; or, when turning fine pivots by means of a small loose pulley runner, cotton does well.

A hand wheel, like that used by some for a watch lathe, can be used to drive a pair of turns, and so get a continuous motion. For this purpose Swiss turns are now often provided with a loose pulley runner. This will be found a great improvement upon the bow for many purposes; but, however the turns are improved, they will never equal the watch lathe in capacity for rapid work. No watchmaker worthy of the name should be without one.

The work in a watch lathe is held more solidly and rotated at an even speed. This enables a heavier cut to be taken, and on all work, except pivots, the back edges of the graver can be used for cutting in the same way as an engineer uses a hand tool. A graver held thus (Fig. 40) rests firmer upon its bed, and takes off a good continuous shaving.

CHAPTER VI.

MAKING SMALL TOOLS.

In former times apprentices made the greater part of their own small tools, and it was very good practice in filing and turning to do so. The tools so made were generally of more lasting quality than modern bought ones. No one now thinks of making his own tweezers, screwdrivers, broaches, and turning arbors, because such things are purchased so cheaply and are so good. But the same cannot yet be said of *all* tool-shop productions.

Drills.—It is necessary occasionally to make a drill, the process being as follows: A piece of steel wire (a broken broach is good if softened) is filed flat on the end and held in a pin vice. It is rested on the edge of the anvil or stake, and "spread" cold with a light blow or two of a watch hammer, as at A (Fig. 54). B shows it spread. It is then shaped by a file like a finished drill, as at C, hardened in oil, brightened with an emery buff or on an oilstone, and tempered to a pale straw for ordinary work, or left hard if for tempered steel. A drill so small that it cools before reaching the oil is hardened by "flirting," as described under "the hardening of steel."

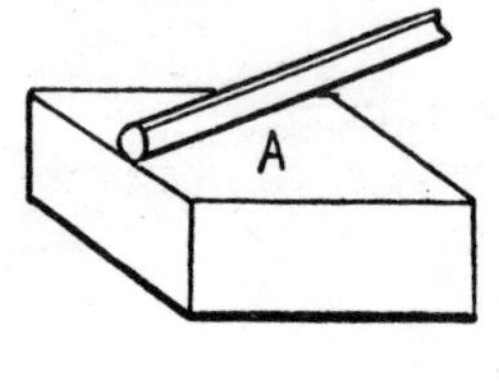

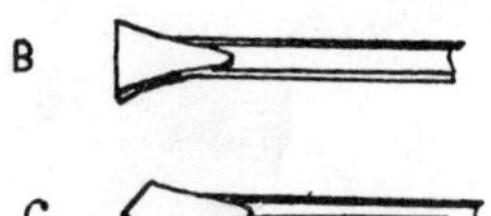

Fig. 54.—Making a Drill.

Taps.—To make a tap, take a piece of well softened steel wire, taper it gradually to just enter the hole of the screw plate; work it in very carefully with plenty of oil until a full thread is cut for a sufficient distance, then lay it in a groove on the boxwood block and file three flats upon it, making it triangular for its whole length, slightly tapered all along and a little more tapered towards the extreme end, where the thread should just disappear. A tap should be hardened in oil and tempered to a full straw. Each flat is smoothed on the oil-stone to give sharp cutting edges. Finally the number of the screw-plate hole is filed in Roman characters on its back end. In small sizes triangular taps cut much more freely than square ones. This is a "taper tap." A "plug tap" has the full thread left right up to the end, and is to follow the taper tap when cutting a thread in a shallow hole in which the screw is required to reach to the bottom.

Punches.—A centre punch should be *turned* from good tool steel rod $\frac{1}{8}$ in. thick, or from a large broken broach well softened. Its shank should be soft, but the point hardened in oil and left hard. Fig. 55 shows the shape. A gentle taper enables the exact position of the punch to be seen when placed on small work, and the sudden short taper at the point gives strength.

FIG. 55.—Centre Punch.

A pivot shoulder punch is for placing over a balance-staff coned pivot to drive on a roller or for other operations. It is made of square steel $\frac{1}{8}$ in. square, and may be made from an old file handle. It has a hole drilled through near to one end, leaving the face thin. A pivot drill is then put through the centre into the big hole and chamfered out a little. The coned pivot operated upon passes through this small hole in the punch face, and the chamfered part rests upon the pivot shoulder and does not injure it. Fig. 56 shows one. These punches are left soft, so as not to mark the pivots.

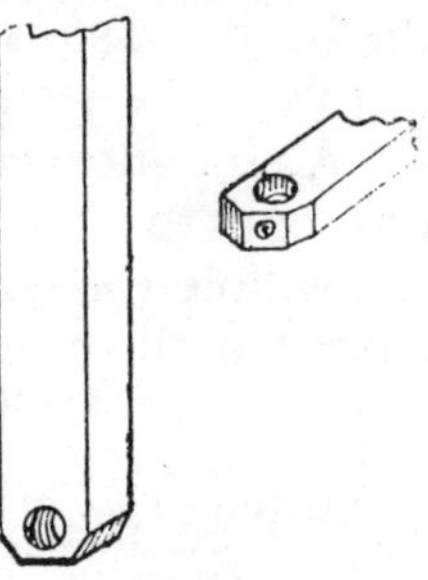

FIG. 56.—Cone Pivot Punch.

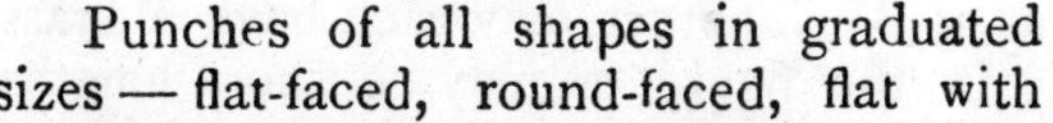

Punches of all shapes in graduated sizes — flat-faced, round-faced, flat with central hole, etc.—can all be made from steel rod, turned up and drilled. The shanks of all should be left soft, and the faces only hardened and polished. No tempering is needed. The tops of punches where the hammer strikes them should be slightly rounded.

Screwdrivers are so cheap and good that ordinary ones are best bought, but a useful one for specially small screws—smaller than jewel screws—can be made from an ordinary needle, sharpened on the oilstone to a chisel edge, and soft soldered into a length of small brass bushing wire for a handle. Such a screwdriver is useful for turning the screws that fasten some white dials on from the front, and does not chip the enamel. To soft solder the needle in the brass, some killed acid will be wanted as a flux. Killed acid is spirits of salt in which zinc cuttings have been dissolved until the acid will dissolve no more. A little acid is put in the hole, some is put on the needle, and the needle inserted in the hole. A small piece of tinman's solder is placed at the joint, and the whole held over the flame of a spirit lamp. When hot, the solder will run and flow in the hole, adhering where the acid has run. While hot, dip it in water and wash well to remove all traces of acid, which would otherwise rust the steel.

Many watchmakers use oil to prevent rusting after soft soldering. This is no good at all. The *only* safe method is to thoroughly wash in water and dry off quickly. Steel parts treated thus never rust at the joint or where the acid has been.

Countersinks.—A set of chisel countersinks for sinking the heads of square-headed screws is very useful. They should run from the jewel screw size to that of pillar screw heads.

A straight piece of steel wire is taken, and a hole drilled up one end to take a central pin of such a size as to enter the screw hole easily. It is then put in the lathe or turns, and a short length turned down to the required size of the sink it is to cut, as at A (Fig. 57). This is then filed to a thin chisel edge like a screwdriver, and hardened and tempered as a drill. It is then sharpened well on the oilstone, and a truly filed tempered steel pin driven in the central hole and rounded up on the end smooth, as at B. The set may each have a brass ferrule driven on, as in the figure, or may be short lengths fitted to a drill stock, or may be made to fit a lathe chuck.

The usual set of circular countersinks sold are very good

for work of such a size as to admit of their use. For making a delicate dot to start a drill, a small *pointed* chamfering tool

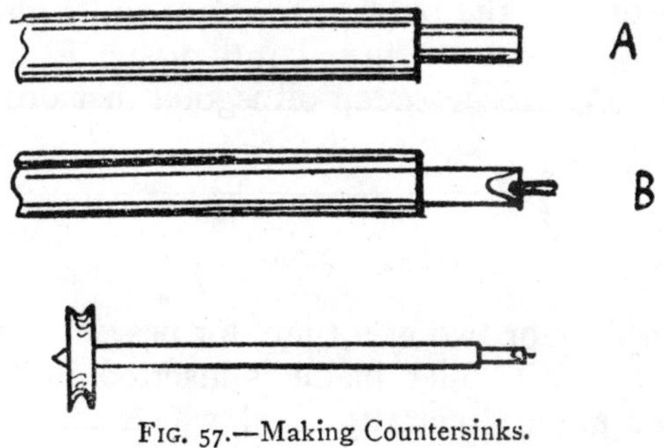

FIG. 57.—Making Countersinks.

is useful, and can be made from a piece of pinion wire or steel wire $\frac{1}{8}$ in. thick and about $2\frac{1}{2}$ in. long. One end is turned down parallel and the point sharpened, as shown in Fig. 58 at A, by being rubbed down or filed to a triangular point having three flats and three cutting edges. This is hardened in oil and left hard, and must be kept quite sharp for use.

A similar sized countersink or chamfering tool, with a semi-circular edge, sharpened like a round-bladed chisel, is useful for just taking the edge or burr off a pivot hole or other small hole.

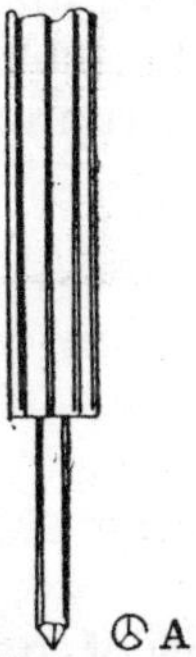

FIG. 58—Pointed Chamfering Tool.

Joint Pusher.—This is for pushing out joint pins. A piece of steel rod about half as thick again as the average large joint pin is set in a handle made of $\frac{3}{16}$ in. round brass rod, by drilling up the rod and driving and soft soldering the steel into it. The steel is gently tapered down to a convenient size for joint pins, and the end is left *flat*. The brass handle has a circular turned brass cap riveted on for applying pressure by hand.

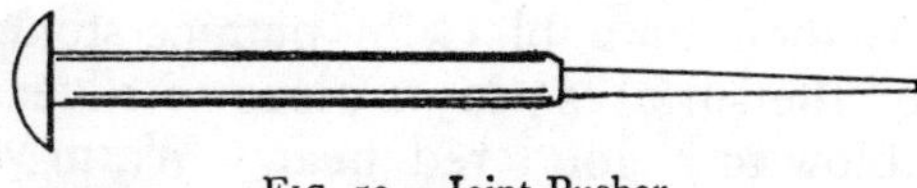

FIG. 59.—Joint Pusher.

The steel portion is hardened and tempered blue. The whole may be 4 in. long and of the proportions shown in Fig. 59.

The Oiler is to apply the watch oil to the pivots and other parts requiring it. It is made from a 4-in. piece of steel wire, spread like a drill at the blade, very thin and rounded on the edge. The handle can be turned into a ring for convenience. Fig. 60 shows one. The watch oil is kept in its original bottle

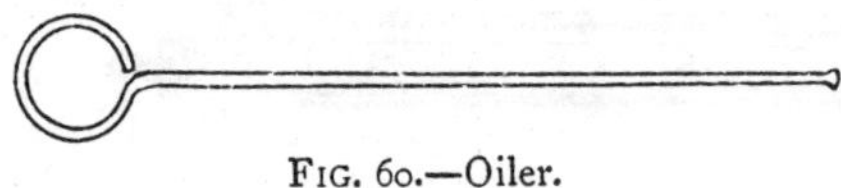

FIG. 60.—Oiler.

and put out a drop or two at a time, for use, into an "oil cup" having a cover. The oiler blade is inserted into the oil cup and takes up a minute quantity of oil, which can be transferred to the pivot holes, etc., by just touching them. A drop or two from the original stock bottle is enough to oil a dozen watches.

A brass "blueing slip" is an oblong piece of thin brass sheet, about $1\frac{1}{4}$ in. × $\frac{5}{8}$ in. To it is "hard soldered" an iron wire handle—iron, because it can be held in the fingers, heat not running up iron like brass. Fig. 61 shows the appliance. A

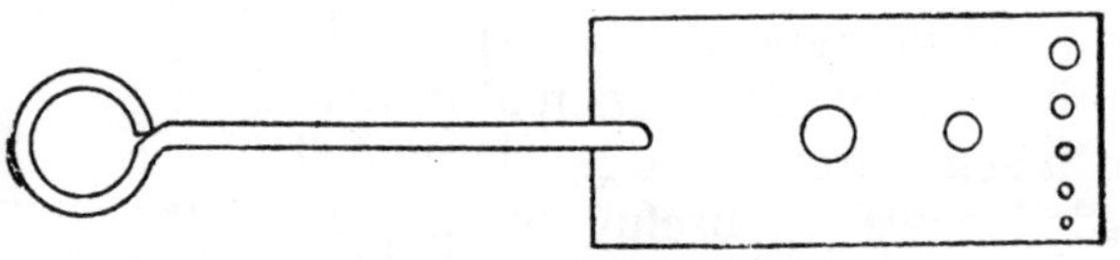

FIG. 61.—Blueing Slip.

row of graduated holes is drilled across the end to accommodate screws for blueing, a larger hole or two being behind them to take the pipes of hour hands so that the hand may lie flat on the brass. Another similar slip may be made with no holes, for blueing flat pieces of steel. This one may also have a groove filed across its centre, so that pieces of round steel wire will not roll off while heating.

To hard solder the handles on, silver solder is used, and borax and water rubbed to a paste is used as a flux. File the parts to fit; lay them on a block of pumice stone in position, having boraxed the surfaces; lay a piece of silver solder along the join, and blow to a good red heat. Warm very gently at first, as the borax will boil up and may displace the parts; then raise gradually to running heat, when the solder will flow in.

Brass spelter may also be used for hard soldering, in the same way as silver solder; but it runs at a greater heat, and does not flow so nicely.

A pair of brass tweezers, for handling the polished steel parts of new watches, are very useful, and may be made from two pieces of thin brass sheet thinned by filing and riveted together. Before riveting they should be well hardened by hammering. Brass tweezers, being soft, will require constant filing up of the points to keep them in good condition. Tweezers that have become slightly round on the edges with wear are apt to shoot off small screws and other parts.

CHAPTER VII.

CLEANING WATCHES.

WATCHES require cleaning for two reasons. They may be dirty through dust having entered and settled upon the moving parts, or the oil may have dried up around the pivots and become sticky. Thus a watch may want cleaning although it is not really dirty at all and may not even have been worn.

To cleanse dirt and sticky oil from watches, they must be taken apart and the pieces put in benzine or petrol, both of which are solvents of grease of all kinds. They are then brushed clean with a watch brush charged with a little dry chalk.

Key-wind Geneva Watches.—To take a watch apart, the first thing is to take it out of its case. Watches are fastened in their cases in many different ways. Taking key-wind Geneva watches first, they are generally held in by a short pin or pins on one side, just entering the edge of the case, and by a dog screw, sometimes two dog screws, on the other side. The dog screws can be seen when the case back is opened, and may be on the top plate, overlapping the case edge a little, or they may be on the bottom plate. Fig. 62 shows a dog screw. Generally they are cut away as shown, so that a half-turn enables the movement to be taken out of its case; but sometimes they are not so cut, and have to be entirely removed. In replacing a movement fastened in like this, the pins are first got into position, and then the movement is pressed down to its seating and the dog screws turned. All watches, without exception, are put in from the case front.

FIG. 62.—Dog Screw.

When out of its case, remove the hands by drawing them off with cutting nippers; when there is no room to insert the edges between the minute hand and the hour hand, insert the edges under the hour hand and draw both off together. Draw the seconds hand off in the same manner. Special pliers and tweezers are made for getting hands off, and can be used by those whose stock of tools is not limited; they are safer and easier to use than cutting nippers. Using a pocket-knife to lever them off is the worst method, and should not be practised.

Remove the dial. If a white enamel dial, it may be fastened by pins through its feet, in which case they are removed with the aid of the small blade of a pocket-knife; or its feet may be slotted and held by two dog screws, which require half a turn *in*, and the dial can be lifted off. Do not force an enamel dial; it will not spring, but crack. Gold or silver dials, and some white ones, are snapped on like box lids over the top edge of the watch plate, and can be removed by levering up at the edge with a pocket-knife.

The dial off, the motion wheels will be seen. The hour wheel which carries the hour hand can be lifted off; the minute wheel can also be lifted off its stud. The cannon pinion is pushed friction tight on to the centre arbor, which carries the minute hand and passes through the hollow centre pinion.

If the arbor projects through the cannon pinion, hold the watch in the hand, and give a smart tap with the hammer and it will probably go through. This will at least loosen it so that by grasping the square at the back in a pair of cutting nippers and the cannon pinion "pipe" in a pair of brass-nosed pliers it can be twisted off. If the arbor end is flush with the cannon pinion top, twist it off as above described; or if too tight for that, lay the watch over a hole in the graduated stake, and with a small hard steel punch tap the arbor through.

If the watch is a horizontal, that is, having a cylinder escapement, before the balance is removed the mainspring must be let down. This is easily done by placing a key on the winding square and holding the click back, letting the key run slowly back in the fingers. In ¾-plate watches the click is covered by a cap fastened by three screws, which must be first removed; the balance-cock screw can then be removed and the cock levered up. It will come away together with the balance and hairspring. A little shaking will free it from the scape-wheel teeth. Unscrew the scape cock and remove the scape wheel. Take off the top plate and remove the remaining wheels. In a bar movement each wheel is held by its separate bar or cock, and care should be taken not to mix the screws. In a lever watch the balance may be removed before letting the mainspring down, but this should be done before removing the pallets and scape wheel.

The benzine is kept in a glass jar with an airtight cover, and must be kept well away from a flame. Into this jar place the plates and wheels, all except the barrel and balance. Let them soak while the barrel and mainspring are taken in hand.

The barrel of a $\frac{3}{4}$ plate is the simplest. Underneath will be found the "stopwork." Hold the winding square in a pair of sliding tongs and prise off the centre steel piece or stop finger. This may be only pushed on friction tight or may be pinned through. Remove from the tongs, and with a screw-driver prise off the barrel cover. This is snapped in a groove in the barrel edge. Then remove the steel arbor. Put arbor and cover in the benzine. Do not pull the spring out of the barrel if it looks all right, but wipe it out well with dry tissue paper, free from oil and dirt. Sharpen a watch peg and clean out the centre hole, twisting the peg round and scraping it clean again until the hole no longer marks the peg. Take the arbor and cover out of the benzine, and, holding them in tissue paper, brush them clean and dry. Peg out the hole in barrel cover.

For re-oiling the barrel and mainspring, a bottle of the best French clock oil should be used. Having brushed the barrel clean and its teeth also, take the arbor and oil its top pivot, where it works in the barrel, with the clock oil. Place the arbor in position and turn it round, seeing that it is hooked properly in the eye of the spring. Apply oil to its bottom pivot, where the cover comes, and some oil to the coils of the spring. Then snap the cover on with the fingers or by pressing it against the wooden edge of the work board. Never use pliers for this. The cover goes on in one position only. There may be a small projecting pin in the barrel groove that goes in a small notch in the cover edge; there may be a dot half on the cover edge and half on the barrel edge; there may be a dot on the cover near the edge and another on the side of the barrel to match it; or the slot for removing the cover may merely have to go next the dot on the barrel side.

When together, hold the square in the sliding tongs and the barrel in the fingers, and feel if the arbor has "endshake" and is quite free. Then wind the spring up to the top and count its turns. As it runs back slowly, feel if it catches or binds inside the barrel. Suppose the spring gives five turns. The stopwork allows four to be used. There will be a turn to spare, so divide this between the top and the bottom. "Set up" the spring half a turn. To do this, wind it up half a turn and hold it there while the stop finger is replaced in the position shown in Fig. 63, seeing that the star wheel A is as shown in the figure. Upon releasing the barrel, the stopwork will then hold the spring wound up half a turn. Wind up the spring to the top to see that all is right. There should be half a turn of spring to spare, thus preventing the spring tearing at its hook in the barrel.

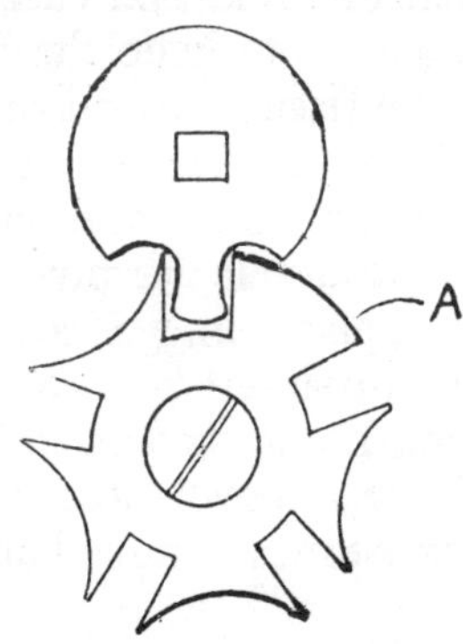

FIG. 63.—Geneva Stopwork.

The barrel of a "bar" movement is more complicated. First take off the top cap or "chapeau" that covers the arbor. Remove the stopwork and barrel cover. Take hold of the centre coil of the mainspring with fine pliers and ease it off its hook. Then pull it out of the barrel. Place a key (birch, black-handled universal keys are always used by watchmakers) on the square, and grasp the arbor inside the barrel with pliers and unscrew it. The barrel, arbor, and bar will then come apart. Put all in the benzine except the mainspring, and clean and peg out. Wipe the mainspring with tissue paper free from grease. To put together, place the arbor in the key and oil where the bar comes and where the barrel turns. Put on the bar and the barrel. Screw on the arbor inside. Place the eye of the mainspring upon the arbor and hook it, and, holding and guiding it with the fingers, wind it in the barrel with the key. When in it will generally hook itself. If not, press the spring well down and wind it up a turn or two. This will hook it. Oil it and place on cover and stopwork as before described. Take out the wheels from the benzine, and, holding them in tissue paper, brush them clean and dry. See that the teeth are clear, and peg out the pinion leaves clean last of all. Clean the cocks, bars, and plates in the same manner.

The watch brush should be soft and thick, and just lightly rubbed on a billiard chalk before cleaning each piece. The chalk is to keep the brush clean as much as to clean the parts. The tissue paper keeps the fingers from soiling the parts and keeps the brush from rubbing the fingers. The constant contact with the tissue paper cleans the brush.

Peg out the pivot holes very carefully with a fine peg point, scraped thin. If a peg point breaks in a pivot hole, sharpen a fine-pointed peg to a somewhat blunt angle, so as to be both sharp and strong. With this push out the broken peg from the back. Jewel holes covered by "endstones" must have the endstones removed and rubbed clean on a piece of wash-leather.

FIG. 64.—Watch Movement Holder.

The holes must be pegged and the endstones replaced. When all are cleaned, screw the bottom plate, or "pillar plate," in a watch holder (Fig. 64). This avoids handling the clean plate, and leaves both hands at liberty for working upon it.

If a $\frac{3}{4}$-plate watch, put the third and fourth wheels in place, then put *watch oil* on the centre wheel and barrel pivots and put them in place. Put on the top plate. Put a piece of tissue paper over it and press lightly on it with the fingers, slipping the top pivots of the wheels into position with a fine pair of tweezers. Use no force in this operation. When right the plate will drop down in position.

Watch oil is not taken direct from the bottle for oiling watches. Such a proceeding would soon render the oil useless.

An oil "cup," with a cover, is used for the workboard. Into the cup about two drops of oil are placed, by dipping the oiler into the stock bottle and transferring the drop into the cup. To apply oil to a pivot, the oiler blade is dipped into the cup and a minute quantity of oil transferred to the pivot. Two drops in the cup should be sufficient for a day's work.

If a bar watch, screw on the barrel bar; then put in the centre wheel and bar; then the third and fourth wheels and their cocks.

Put in the scape wheel and pallets if a lever, placing just a little oil on each pallet face before putting them in.

Each wheel should be tried carefully, to see if it has endshake, as the watch is put together. Every wheel and arbor in a watch *must* have endshake, or they will bind and stop the watch. The amount of shake or lift must be enough to see with an eye-glass, but should not be excessive. An amount of shake equal to half a pivot length is far too much. Fig. 65 shows a fair amount in proportion to a pivot. When wheels are held by cocks, the endshakes can be regulated by inserting small pieces of tissue paper, or thicker paper if necessary, under either the front or back end of the cock. More endshake can be given to wheels under a $\frac{3}{4}$ plate by putting a paper washer upon the pillar top under the plate. The barrel arbor alone need have no endshake between the plates.

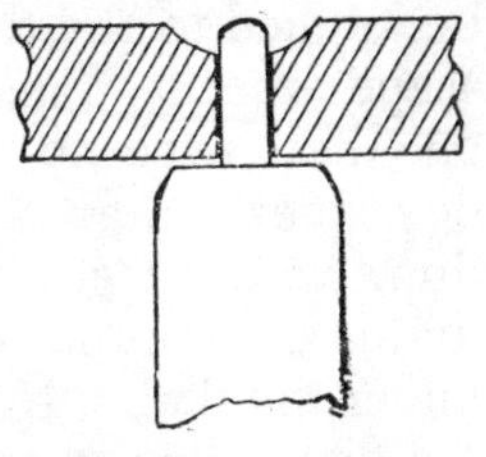

FIG. 65.—A Correct Endshake.

The set-hands arbor can be cleaned and put in, and the cannon pinion pushed on home. This may require a tap with the hammer and a suitable hollow punch, passing over the cannon pinion pipe and resting on the leaves.

The balance and balance cock, etc., can now be taken in hand. The hairspring stud must be first removed from the cock. Usually the stud is pushed friction tight into a hole in the cock. In this case the cock edge may be rested on a stake and the stud pushed through from the top. When loosened thus, complete its removal with tweezers from underneath. The cock can then be taken apart and cleaned with benzine, the jewel hole and endstone cleaned and replaced, together with the regulator index. The hairspring should not be removed from the balance, but all may be dipped into the benzine. It can be dabbed, hairspring downwards, on paper to remove the benzine, and given a minute to dry off. The cylinder should be cleaned out inside with a peg. The pivots and balance rim can be cleaned by pith.

Place a small quantity of watch oil in the jewel hole, replace the hairspring stud, pressing it well down in place, and see that the hairspring goes well between the curb pins in the regulator without pressing hard against either of them. Put a little oil on each scape-wheel tooth and in the lower balance jewel hole and put balance in. Screw the cock down very carefully, seeing that the pivots go into the holes and that the balance is free before screwing quite tight. The endshake of the balance is very particular. A little, but only just perceptible, is wanted. It can be regulated with tissue paper, as before described. The watch can then be wound, and oil can be applied to the top and bottom pivots of the train wheels and scape wheel. The minute wheel and hour wheel can be put on (these need no oil) and the dial replaced. If a dial held by dog screws, draw the dog screws outwards to tighten, so as to draw *up* the feet; this prevents the dial rattling. When the dial is on, see that the hour wheel has a little endshake, or lift, under the dial. If too much, put paper collets over it under the dial. Put on the seconds hand, seeing that it lies close to the dial and flat, but does not touch it anywhere. Put on the hour hand, seeing that it has a little shake and is not bound against the dial, also that it just clears the seconds hand. Put on the minute hand, and, resting the set-hand square on a steel stake, tap it on with the hammer. See that the minute hand is well clear of the hour hand, but that it does not stick up enough to touch the glass.

Dust out the case and replace the movement.

The various parts of this movement have not been illustrated, as the figures and accompanying descriptions in the introductory chapter will be found sufficient.

Particular care must be taken that oil is applied to every pivot. Neglect of this leads to certain disaster. The friction causes rapid wear and rust. The rust powder, in itself a polishing and cutting powder, rapidly cuts the pivot away, until it disappears entirely, leaving only a mass of rust in the pivot hole.

Keyless Watches.—This description of the method of cleaning and putting a Geneva watch together applies to key-wind and keyless watches alike as far as it goes. But keyless watches require some further attention as well. The first difficulty with a keyless Geneva watch is to get it out of its case. They are fastened in in various ways. Dog screws like those shown in Fig. 62 actually hold the movement; but before it can be removed the winding stem, and sometimes the set-hand side push piece, must be taken out.

In some watches a small screw is found in the case pendant, and removing this enables the button and winding stem to be pulled straight out. In most a small steel screw either at A or B, Fig. 66, has to be withdrawn a turn or two before the

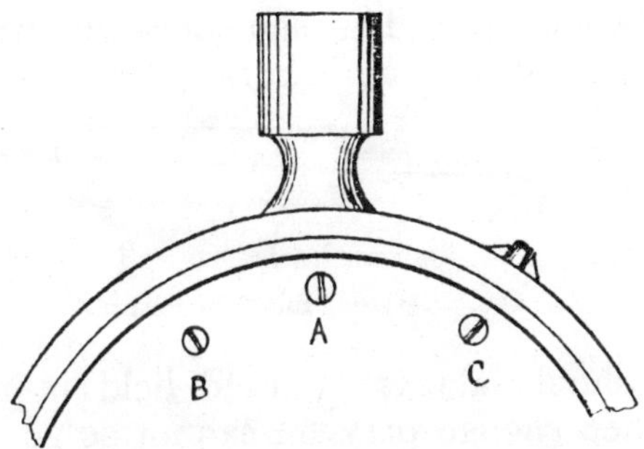

Fig. 66.—Position of Pendant Screws.

stem can be drawn out. If there is also a screw at C opposite the side push piece, draw it out and remove the push piece; but in most watches the movements will come out without removing this, and no screw will be found at C.

Occasionally the screw fastening the winding stem in is underneath the dial, and the hands and dial must be first removed to get at it. A very few watches (mostly by "Patek Phillippe") have the winding stem fastened in such a way that to get the watch out of its case the bar holding the winding wheels in the watch must be taken out and several other parts also, as well as the hands and dial, to remove it.

A keyless watch should have the mainspring "let down" before it is taken out of its case. This is done by holding the winding button in the fingers, while the click is held back, and allowing it to slowly run back.

There are many kinds of keyless work, and these will be described in a chapter to themselves. In cleaning the watch, all keyless wheels and parts should be taken off and cleaned in the benzine. In putting together, all want well oiling where they rub against each other or the watch plates or bars. In taking them apart, it is well to remember that top keyless winding wheels are frequently fixed by a large central *left-handed* screw, which, of course, requires turning to the right to unscrew it.

English Watches.—English watches may be full plate or $\frac{3}{4}$ plate, and keyless or key-wind.

Taking full-plate, key-wind fusee watches first, the movements are generally fastened in their cases by means of the joint pin at the Fig. 12. This should be removed by a "joint

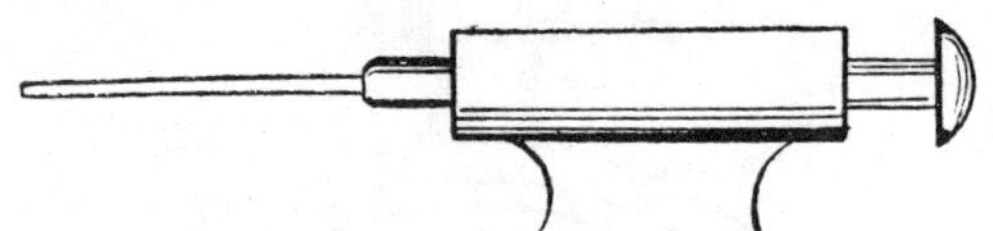

FIG. 67.—Using the Joint Pusher.

pusher." It should always be used held in the hand when possible. When enough pressure cannot be got in this way, it may be screwed fairly tight in the vice, as shown in Fig 67; the watch is held up to its point, and a smart blow given to its head with the hammer. This will dislodge the most stubborn joint pin. Before resorting to this method the balance had better be removed, for fear of injury to its pivots.

When out of its case, remove the hands, take out the balance cock and balance, and remove the dial. Dial pins may be removed by the aid of a pocket-knife, levering them forward. Sometimes a pin cannot be got out in this way because none of it projects. In such a case it must be got out by applying pressure from the back, or inner end, by inserting a screwdriver blade and pushing, or in stubborn cases using a steel draw-hook to pull them forward. Such a hook is shown in Fig. 68. It is inserted behind the inner end of the pin and pulled forward.

FIG. 68.—Pin Draw Hook.

The mainspring must then be let down. It is held by a ratchet and click under the pillar plate. Loosen the click screw half a turn, place a key upon the square, and let it down. Sometimes there is not enough square for a key to hold. Then screw a pin vice on to the fusee winding square, screw the pin vice in the board vice, like Fig. 69, and, holding the movement firmly in the hand, take off the under bar that holds the third and fourth wheels. Take out the third wheel, replace the bar and screw it down, then let the movement slowly turn round in the hand as the mainspring unwinds itself.

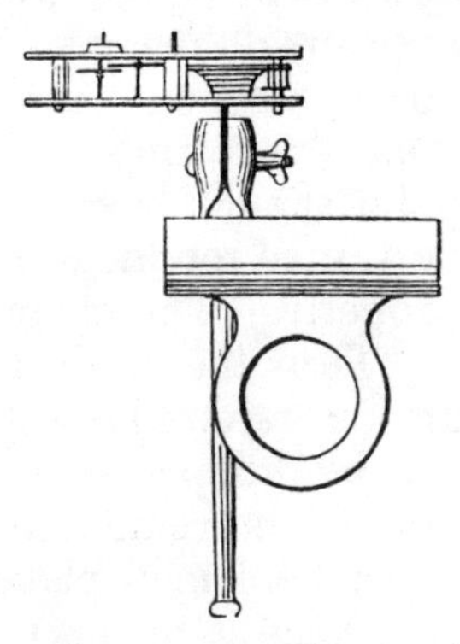

FIG. 69.—Letting down a Mainspring.

When let down, take off the barrel bar or name bar and remove the barrel, having unhooked the chain first. Then take out the four pillar pins that hold the top plate, and gently raise the plate a little to enable the lever to be lifted out of its pivot holes and withdrawn. If the plate is taken off before taking this precaution, the lower pivot of the pallet staff will be bent or broken, as the lever will be caught by the potance. The potance is the brass cock screwed to the under side of the top plate to carry the lower pivot of the balance.

Then lift off the top plate and remove the fusee and wheels, etc. The cannon pinion can be removed from the centre arbor by grasping its square in a pair of cutting nippers and pulling or twisting, holding the centre wheel tight in the mean time.

Plates, wheels, bars, etc., can be put in benzine. The barrel and mainspring can be cleaned as in a Geneva watch. The chain can be wiped in slightly oiled tissue paper. The fusee, if the clickwork inside it seems sound, need not be taken apart, but merely brushed clean, and *not* put in benzine. If the fusee winds too hard, run a little oil between the edge of the steel maintaining ratchet and the brass fusee body.

In putting together, put the plate in the holder, place the third wheel in position, oil the bottom pivot of the centre wheel and place that in position, oil the fusee pivots top and bottom, and put the fusee in. Put the maintaining detent in place the wrong way round, as it then stands up more easily and can be turned right afterwards. Put in the fourth and scape wheels; then put on the top plate, and before it is got down to its place, introduce the lever and get it in position with tweezers; then get the pivots in their holes and the top plate down, pinning it on. The barrel goes in next and its bar, then the cannon pinion. The maintaining detent (whose point should have been filed up sharp before putting in) can be turned round, and should be examined to see that it engages properly. The chain now has to be put on.

Turn the fusee and barrel round so that the chain holes are outward. Hold the movement vertically so that the chain can be dropped clear through it from fusee to barrel. Then hold the movement in the left hand, and with tweezers insert the barrel hook in its place. Place the thumb of the left hand upon it to keep it in place, and putting a key on the barrel square, wind the chain upon it until it is nearly all on. Hook the

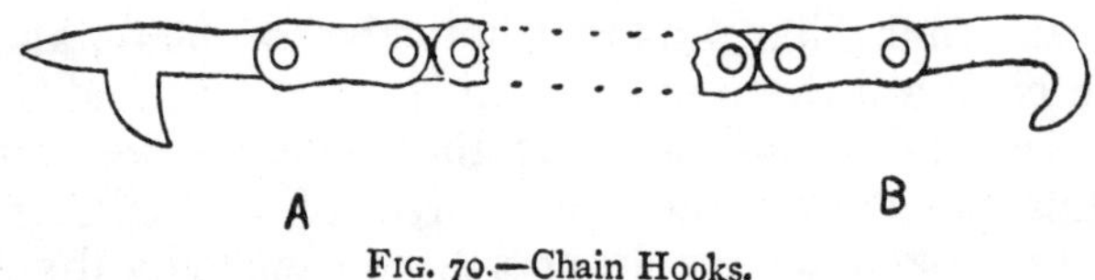

FIG. 70.—Chain Hooks.

other end in the fusee and turn the barrel round a little further to draw the chain tight. Now place on the barrel ratchet and "set up" the spring half a turn, or three-quarters if the spring will allow, and screw the click tight.

In setting up the spring the want of a third hand is often felt to put the click into the ratchet teeth. This want can be supplied by holding a long peg in the mouth and using it to manipulate the click. Fig. 4, p. 3, shows the arrangement of a barrel, chain, and fusee. Fig. 70 shows the two ends of a fusee chain. A is the barrel hook, B is the fusee hook.

When all is wound upon the barrel and set up, the watch can be carefully wound up. As this is done, see the chain runs straight to the fusee and does not drag aslant, or it will get out of its groove.

Now clean the balance and cock, oil the balance holes, the top pivot holes of the centre, third, fourth, and scape wheels, and lever (a very little indeed to the lever). If much is put to the lever it may run between the lever and plate and clog. Put in the balance. In a watch in which the hairspring is above the balance (oversprung) the stud will simply be screwed in place. In an undersprung watch the spring will require re-pinning in its stud, which is a fixture on the plate. And it must be pinned in at exactly the right place to set the watch "in beat." To test for this, wedge the fourth wheel with a peg or a pivot broach, and if in beat, when the balance is at rest the ruby pin in the roller will be in the lever notch, and the lever will stand midway between its banking pins. Adjust the hairspring until this is so, and then see that it passes between the index or regulator "curb" pins centrally and plays between them evenly, especially when the regulator is at "slow." See also that it lies flat and does not touch the balance or the plate.

Apply a little oil to the points of the scape-wheel teeth, oil the bottom pivots, put on the motion work, dial and hands, and put in the case again. See particularly that the dial pins fit well and go in tight; also that none of them touch the wheels, etc.

The fusee is out of date in lever watches, and not now made except in special cases. Most modern English full-plate watches and all American ones have "going barrels" like Geneva watches.

They are generally arranged so that the barrel bar can be taken off and the barrel removed without taking off the top plate. They all should be so made. The mainsprings of these watches can be let down by a key on the winding square, while the click is held back. In putting them together, it is advisable to put the barrel in place with the rest of the wheel-work before putting on the top plate. The barrel then steadies the plate and helps to hold it in the right position while getting the lever in. Otherwise these watches are cleaned in the same manner as fusee watches.

English and American $\frac{3}{4}$-plate key-wind watches will present no special difficulties. The principal difference is that these watches have the top plate cut away; and the balance and escapement are held by separate cocks screwed to the pillar plate. This has several advantages. First, it secures a flatter watch; second, the balance is less likely to be crushed and injured through accident; third, the hands can be set from the back; fourth, the escapement can be removed separately.

Besides these, other forms of English full-plate watches will be met with—the verge, the cylinder, and the duplex more especially, and a few pocket chronometers.

Verge Watches, the oldest form of all, now not made, are rapidly dying out. Still, it may be many years before they are all worn out or disused, and the country watchmaker still has them with him.

A verge resembles a lever up to the fourth wheel. The fourth wheel is a "crown wheel," and drives the scape pinion and wheel, which is carried under the top plate by a brass "follower," and the potance, as in Fig. 71. A is the potance, B the "follower," and C the scape wheel. The verge or axis of the balance generally runs in brass holes top and bottom.

In cleaning these watches, do not remove the follower to take out the scape wheel, but unscrew the potance. This is less likely to disturb the escapement. The dead brass holes in follower, potance, and balance cock must be very carefully cleaned by pegging out, and care taken not to break a peg in. In putting together, apply oil to the lower verge hole first and to the scape pivot hole in the potance, as these cannot be got at afterwards. Then put the scape wheel in position and oil the follower hole. The train wheels and top plate can then be put on. Put on the chain before putting the verge in. Set in beat by seeing that it cannot be stopped on either pallet, but starts off immediately it is released. If the balance cock has to be removed for any purpose after the watch is wound up, the fourth wheel must be wedged as a precaution. If not, the train may run and damage the scape-wheel teeth. Put no oil on the verge pallets.

FIG. 71.—Verge Scape Wheel.

English Cylinder and Duplex Watches.—The same

remark applies to these watches. The balance cocks must not be loosened while there is any power on the scape wheels. Otherwise these watches are cleaned just as levers. Cylinders require oil inside them and on the scape-wheel teeth. Duplex watches require oil on the ruby roller and on the point of the long impulse pallet.

Pocket Chronometers require great care in handling. The balance should not be removed without first wedging the fourth wheel, or there is danger. The delicate detent and scape wheel should be put in a safe place while the rest of the watch is being cleaned. When putting the watch together, if a full plate, leave the detent out, as it can be put in last thing. To clean the detent, lay it flat on clean paper and hold down the foot with tissue paper. Brush it gently to remove dirt, and with a fine pointed peg clean the locking ruby and the point of contact of the gold spring and detent point.

A chronometer only requires oil to its pivots ; none should be put on the scape-wheel teeth, the locking stone, the gold spring, or the roller pallets.

English Keyless Watches, whether full-plate or $\frac{3}{4}$-plate, will come straight out of their cases without first removing the winding stems.

American Keyless Full-plate Watches generally require the removal of a small set-screw in the case pendant, and the button can be drawn off. This allows the movement to come out.

Three-quarter-plate American Watches are of many patterns. Some are made just as the full plates above described ; others require the loosening of a small screw on the top plate, like some Swiss watches. Although they vary very much in pattern, none will present any special difficulty in taking out of their cases.

The keyless work of all requires cleansing in benzine, and plenty of oil applying to all points where friction occurs.

General Remarks on Cleaning Watches.—In taking a watch apart, everything should be carefully tried before removing, to discover faults. The neglect of this causes a great deal of wasted time. It is better to find a fault when taking apart, and remedy it before cleaning, than to find it when the watch is cleaned and put together. Thus, when taking off the hands, see if they have been touching the glass, or if they are too loose or bind against each other. Test the escapement. Try all endshakes, feel all depths, and look for broken or cracked jewel holes before anything is cleaned.

In putting together do not finger cleaned parts. Any spots of black or dirt on the plates, that will not brush off, remove with a peg point. If the gilding has got tarnished, brush it with wet benzine on the brush. If a nickel movement has become tarnished, remove it with spirit on tissue paper, or, in bad cases, with rouge on a cork. Always remember to put oil to jewel holes with endstones before the wheels are put in. Enamel dials can be cleaned with a damp cloth. Gold hands can be laid on a cork and rubbed gently with another cork and rouge powder.

Screws, though they may appear to be interchangeable, are generally not so, and means should be taken to replace them in their correct positions. Some screws are dotted on the heads; if so, they should be placed in the holes dotted to correspond. One of every pair of jewel screws has a dot; this goes in the hole nearest to the dot on the edge of the jewel setting. Pins are seldom interchangeable, and dial pins should be kept apart from pillar pins.

Beginners should be very careful indeed in handling balances and hairsprings. These are the most delicate, as well as the most important, parts of a watch. A very little will bend a hairspring, and the least pressure will damage a balance pivot. Let a balance down gently into its lower pivot hole by its own weight. When putting the cock on and screwing it down, set the balance vibrating, and let it continue to do so while the cock is tightened. Any nipping or pressure upon the pivots will at once show itself by the balance ceasing to vibrate, and attention will be attracted to it before damage is done.

Before putting in the balance of a watch, just see finally if the train is all free by touching the centre wheel or great wheel teeth. If a cylinder, verge, or duplex, the train will run, and its freedom can be judged. If a lever, a little *back* pressure on the fusee or barrel teeth should cause the scape teeth to trip back past the pallets. A lever that will not do this readily is faulty.

A dot upon the centre steel piece of an English rocking bar must be placed towards a corresponding dot on the movement edge, if there is one. If not, put it on with the dot towards the centre wheel. Similarly, an English barrel cover must be put on with the opening next to the dot on the barrel, or, if there is no dot, next to the chain hook hole.

A slight finger mark or smear upon a newly cleaned plate can be removed with clean-cut pith, a clean surface being cut each time the plate is rubbed.

When jewel holes in English or American watches have endstones to cover them, the jewel settings are fitted closely in a sink turned out in the plate, and held in by two small jewel screws. Such jewelling should always be taken out by withdrawing the screws and pushing the jewel hole and endstone out with a flat-ended watch peg. They are both cleaned by rubbing them with the finger on a piece of washleather and pegging out the hole.

American jewelling, and some in machine-made English lever watches, is often fitted so tight in the plate that great force is needed to push it out. In such a case take care that the pressure is applied to the brass settings and not to the jewels themselves. Such tight jewelling can be replaced by pressing in with a flat-cut peg, or by making a flat-ended punch out of pegwood and using the hammer. Some watchmakers keep a few ivory punches for this and similar purposes.

To avoid danger to the hairspring while brushing a roller or cylinder clean, push the roller or cylinder and lower part of the balance staff through tissue paper.

As tweezers wear, the inside edges of the points become smooth and rounded, causing pins and other small parts to "shoot" from them. The remedy is to pass a sharp fine file over their inside faces and thin them a little, bringing the edges up sharp again.

To handle a dial or pillar pin safely, pick it up from the board with tweezers, lay it on the back of the hand, and again take it up in the tweezers. This time the points will get a firmer hold, as the hand back is soft and allows the points to go further over the pin.

CHAPTER VIII.

BARRELS, FUSEES, MAINSPRINGS, AND CHAINS.

Broken Mainsprings.—A mainspring that is broken close to the outer end may be made to do again by punching a fresh hook hole and filing up. First soften the extreme end in the spirit-lamp flame, and hammer it flat. Then, with a mainspring punch, or an ordinary flat-ended round punch, and a graduated steel stake, punch a hole in it. A round hole is as good as any. Broach out the hole aslant, as in Fig. 72, so as to leave a sharp edge outside to hold well under the barrel hook. With a rat-tail file open the broached hole oval and

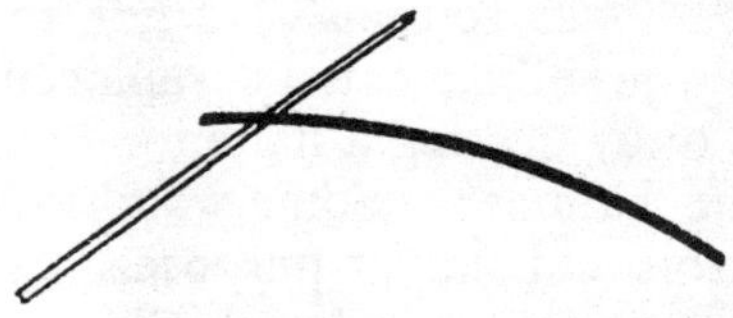

Fig. 72.—Broaching a Hole in a Mainspring.

Fig. 73.—Mainspring End.

make it quite central. Then flat the spring end on both sides by filing to remove all burrs, and shape the end up as in Fig. 73.

This is the usual method of hooking a spring in its barrel. But there are many others. The old way was to rivet a square block of steel into the spring end and cut a corresponding hole in the barrel to receive it. To make such a hook, take a length of oblong drawn "hooking-in" steel of the correct size, and screw it in the vice aslant, as at A (Fig. 74). File it up as at B, leaving a kind of pivot to rivet into the spring. Cut it off as at C. Hold it in a pair of cutting nippers, and file the surface flat where cut off. Then rivet it in a hole punched in the spring ready for it. The spring end in this case should be

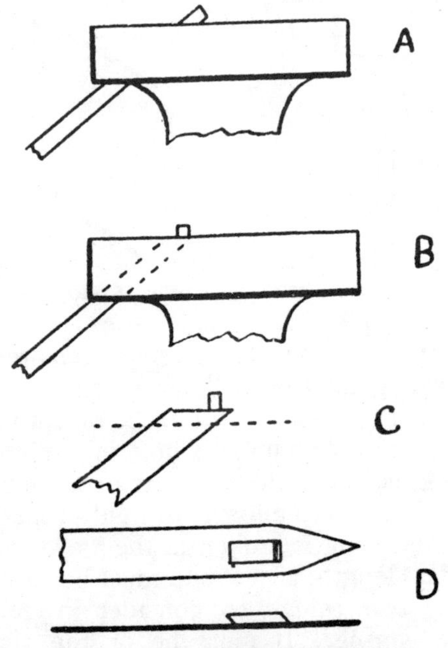

FIG. 74.—Making a Mainspring Block.

filed up as at D, and the pointed end thinned. The block is finally finished off smooth, so as to lie level with the outside surface of the barrel.

Some springs are fastened in with a brace or buckle, like Fig. 75. These braces can be bought ready made, by the gross, or they may be made from a short length of thicker and wider spring. A home-made one requires a hole punched in it and the spring, and a brass rivet to hold all together. The bought braces have a soft steel rivet on them ready for riveting into the spring. The two pivots of the brace go into holes in the barrel bottom and cover, and allow the spring to draw over when wound full up instead of bending. This system is used in all American and many English machine-made watches.

A method used in cheap Swiss watches is shown in Fig. 76,

at A. The extreme end of the spring is softened, and turned back. A short loose piece of spring is slipped in, sharpened at one end to catch on the hook. When the turned back end

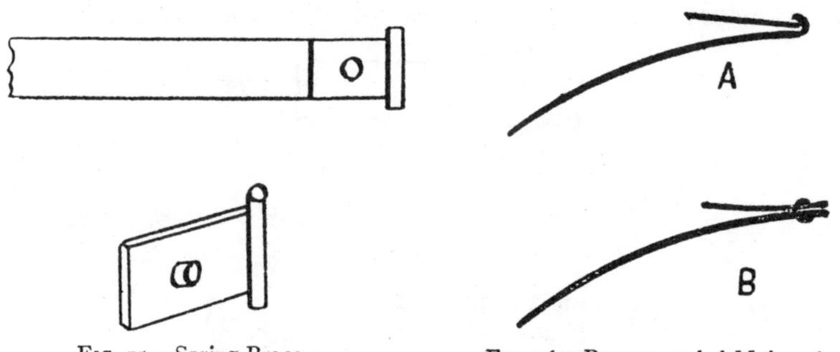

FIG. 75.—Spring Brace.

FIG. 76.—Reverse ended Mainspring.

breaks off, as it often does, the average repairer finds a great difficulty in turning another without breaking it. A way out of the difficulty is to rivet a short piece of spring on, as at B, with a brass rivet. To punch a hole in the end to go direct on to the hook is, as a rule, no good in these watches, as the hook is not of a shape to allow it to hold properly.

When a mainspring is broken in the "eye," or anywhere in the middle of its length, a new one must be fitted. There are two things to consider in selecting a new spring. It must be of the right width and strength. Before pulling the old spring out, see if it is of the correct height or width. It should reach nearly up to the groove in which the cover is snapped. If correct, measure its width in the gauge, and take the packet of springs indicated by the gauge number. To measure the strength, the gauge shown in Fig. 77 is used. Pick out one that matches the old spring. In a mainspring thickness means strength. Two springs of the same thickness are of the same strength. Yet a "dead" spring will not pull like a lively one. In Fig. 78, A shows a good lively spring; B indicates a dead one. A spring that has been once wound in a barrel and comes out like B will not pull a watch as well as a much weaker one that

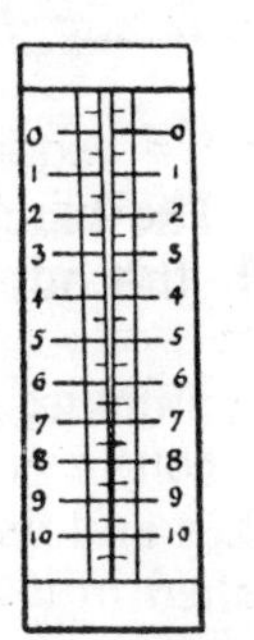

FIG. 77.—Mainspring Strength Gauge.

comes out like A. "Life" in a spring is a question of quality and temper.

Measure the length of the new spring by the old one roughly, leaving it a little too long. Wind it in the barrel with a mainspring winder, and see if it is of the correct length.

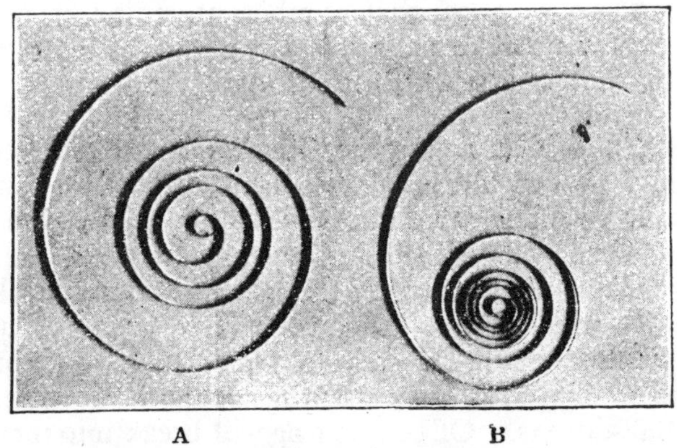

FIG. 78.—Good and Bad Mainsprings.

Fig. 2, p. 2, shows about how full the barrel ought to look. If more or less than this, the spring will not give so many turns. The outer end can be broken off with the fingers until correct. Then soften the end, hook as before described, and finally wind it in.

To wind a spring in with a mainspring winder requires some practice. Hook the eye on the arbor of the winder, wind up a turn or so, holding the spring in the fingers of the left hand. Then place the barrel over as many coils as it will cover, and press it against the end of the winder arbor. Continue winding, guiding the spring in with the fingers and holding the barrel firmly. When all is in, it will hook itself, especially if pressed well down and given a turn or two more.

A spring having a reverse end like Fig. 76, B, or a brace like Fig. 75, is best put in by hand. To do so, put the outer end in first, and coil it in gradually, a little at a time, without straining the spring. Finally, press it well down with a screw-driver, and the brace can be pushed round into its place by the same means. A reverse end will go to its place and catch properly the first time it is wound.

There are many gauges of mainspring widths, mostly based on the millimetre. In one gauge No. 1 is $\frac{1}{10}$ mm., No. 2 $\frac{2}{10}$ mm., and so on. In another gauge No. 1 is 1 mm., No. 2 $1\frac{1}{10}$, and so on. Using either of these gauges, the sliding millimetre gauge with vernier reading to tenths can be used to measure them, and also to measure the depth of a barrel, and see at once what number of spring is required. In posting for a single spring, it is only necessary to enclose an inch of the old one which it is to replace.

Barrels.—If a mainspring breaks in the barrel of a fusee watch, when the spring is nearly wound up, it often expands it so much that the cover is loose, or even makes it so cone-shaped that it will no longer go in the watch. The cause of mainsprings breaking is obscure. They break more often when undisturbed than when being wound up. A change of temperature seems to affect them, and on such occasions watches with broken springs come in to the repairing shop in half-dozens. Often a spring will break into three, four, or up to twenty pieces. Exactly what occurs at the moment of breakage in such cases is an interesting speculation.

When the cover is loose it can be tightened by spreading the edge equally all round with a hammer. Lay the cover, underside uppermost, on a flat, polished, steel stake. Begin at the notch and gently tap it all round with the flat of the hammer until the notch is reached again. Be very careful to hammer *flat* and not dent the cover. Hammer *aslant* to spread the edge, moving the hammer in the direction shown in Fig. 79.

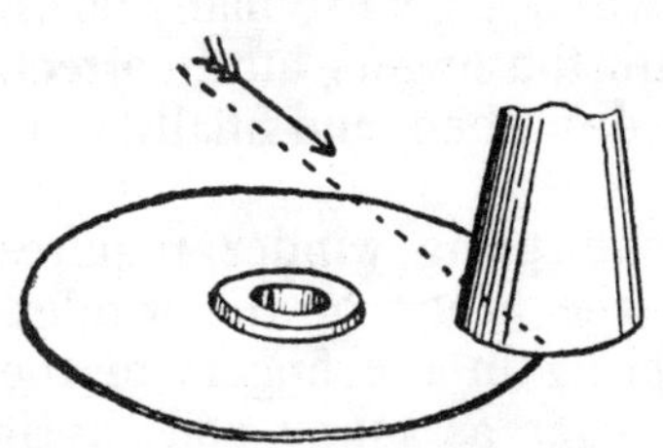

FIG. 79.—Hammering a Barrel Cover.

A barrel spread too much to go in the watch may be closed with care by using a jeweller's ring-stretching tool.

To put in a new barrel hook, drill the barrel centrally, and broach it slanting a little in the direction to hold the spring. Tap the hole with a good full thread. File up a taper hard brass pin, thread it in the screw plate, cut off its end, and file it a little slanting. Then screw it in from outside until it jams tight. If too long inside the barrel, withdraw it and shorten. Leave just sufficient inside the barrel to come well through one coil of the spring. Finally, cut off the pin outside and smooth off. The success of this method depends on jamming the pin tight in the barrel, and to aid this the hole may be tapped a little small by not running the tap more than half its length in, and the pin may be tapped as full as possible.

Very good ready made hooks can be purchased at the material shops. They are made with a long tapered end that can be passed through from the *inside* of the barrel and screwed outwards until tight. The outer tapered part can then be cut off level and smoothed. These hooks are better than those screwed in from the *outside*, as they cannot be forced out by the breaking of a mainspring.

A barrel that is burst cannot be repaired. It is no use soft soldering it up. A new barrel and cover can be bought from the material shop, in the rough. First open out the cover to go tightly on the top pivot of the arbor. This can be done by broaching. Then turn the inside boss, A, down until the shoulder of the pivot comes through the cover, as at B (Fig. 80). Serve the barrel bottom the same and fit it to the lower pivot, thinning the inside central boss until the endshake of the arbor inside the barrel is right. All this turning work is done very easily in a lathe, holding barrel and cover in step chucks and using the slide rest. In the turns the cover and barrel can be put on arbors and turned by hand; or they can be done in a mandrel. To make the chain hook hole, put the barrel in the frame and mark a line on the barrel just below the under edge of the top plate, so that the chain will just touch it if drawn straight away from the hook hole. With the pointed chamfering tool dot two drilling centres, as shown in Fig. 81. Drill hole No. 1 through slanting, as shown at A. Drill hole No. 2 still more slanting, as at B. Then with a pivot broach and oil broach one into the other,

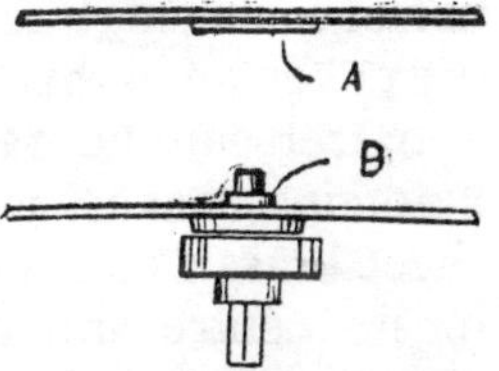

FIG. 80.—Fitting a Barrel Cover.

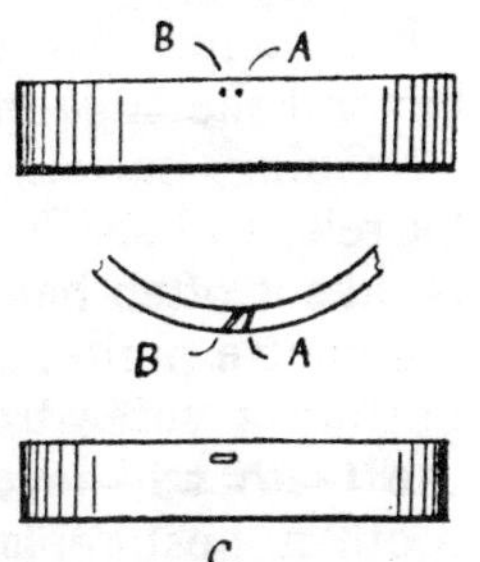

FIG. 81.—Making a Chain Hook Hole.

as shown at C. This makes a good oblong hole, well undercut for the hook. Put in a good sound mainspring hook. For a finish, buff the barrel round with a $\frac{3}{0}$ emery stick, and finally with a dry rouge-on-leather buff. The cover may be served the same, or it may be stoned with water-of-Ayr stone and water by short circular strokes round and round all over it. Barrels are best not gilt, as gilding softens them.

A mainspring that binds in a barrel when running down may sometimes be eased by getting its coils flat. Or, if the cover is a thick one, it may be put in a step chuck and a thin cut taken with the slide rest from the centre boss *nearly* to the edge. If a spring binds much a new one makes the best job.

An arbor that has no endshake in the barrel may be given some by laying the barrel over a hollow turned out in a piece of sheet brass and giving the top of the arbor a blow with the hammer. This slightly bulges the bottom outwards. If a barrel is too high in the watch, it can be lowered by bulging the cover *up* in this manner and the bottom *up* also. Similarly, a barrel can be raised if required. For these operations a "doming stake" is useful. This may be made of a piece of sheet brass 1 in. square and $\frac{3}{16}$ in. thick. A $\frac{1}{8}$ in. hole is drilled in its centre and a shallow coned hollow turned around it, about $\frac{1}{16}$ in. deep and of a diameter equal to the barrel of a lady's watch.

A watch barrel sometimes runs foul of the top bar in an English watch. To remedy, the bar is centred in a mandrel or lathe by the barrel pivot hole, and a cut with the slide rest taken across its underside from near to the centre hole to beyond the edge of the barrel.

Going barrels are generally made stronger than chain barrels, and seldom either stretch or burst; but when a spring breaks it often bends a tooth—the one that happened to be in the centre pinion at the time. These can be straightened by putting a screwdriver blade or a pocket-knife blade between the teeth and levering the bent tooth up. A broken barrel tooth is best replaced by drilling into the barrel, tapping the hole, and screwing in a steel pin, which is then filed up to the length and shape of the other teeth.

A new steel hook can be put in a barrel arbor by drilling a good deep hole into it in a fresh place and driving in a steel pin. This is then filed up and undercut to hold the eye of the mainspring.

Fusees and Chains.—In fusee watches the chain often breaks. To mend a chain, lay it on boxwood and hold the end link down with the finger nail, while a thin-edged pocket-knife is used to open it and start the rivet end. As soon as the side of the end link is raised enough to clearly show the rivet, get a steel needle and flat its end slightly on an oilstone. Screw the needle in a pin vice for a handle, and, placing the chain with the rivet over a small hole in the steel stake, push the rivet through with the needle. If stubborn, tap the pin vice with the hammer. In this way prepare both ends of the chain, leaving a double link to one piece and a single link to the other, as in Fig. 82. For pushing out chain rivets and similar jobs the long pins sold with round glass heads are useful, as they can be used comfortably in the hand.

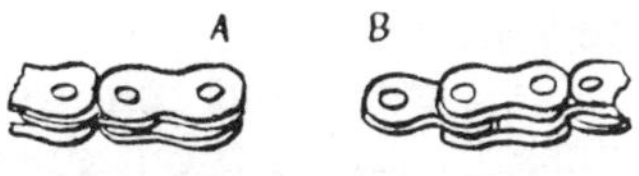

FIG. 82.—Mending a Chain.

File up a tempered steel pin (a needle let down to a blue), and place the two ends of the chain together so that the hooks at the ends come *on the same side.* Insert the pin, still in the pin vice, and tap it in gently over the stake. With nippers cut the end off flush, and with a very fine file, file it flat with the chain. Cut off the other end of the rivet, lay the chain on boxwood, and file both sides flush with the chain. Then lay the chain on a flat steel stake and gently tap the rivet on both sides. The join, when done thus, should be invisible.

If the chain on a partly wound up watch falls flat upon the barrel, take a graver, stick it in the barrel edge, and force the barrel back a little to ease the pressure off the chain. It will then usually fly up into its correct position. If a chain persists in turning over in this manner, reverse the hooks; that is, put the fusee hook on the barrel end and the barrel hook on the fusee end.

When the fusee groove is bad, there is no remedy but sending it to a fusee cutter (the material dealers get such jobs done), together with the chain, and having it re-cut to fit.

The winding work inside a fusee often needs repair. To take it apart, hold the square in a pair of sliding tongs and push out the pin that goes through the steel cap underneath. The cap then comes off, the main wheel and maintaining ratchet following. The ratchet is then seen, and also the clicks. To fit a new ratchet, first lever up the old one with a pocket-knife; it is pinned on with two brass pins. Pick out a new one to pass easily inside the clicks; gauge this by laying it on the maintaining ratchet and observing the clicks. Open out the centre hole by broaching until it goes tight on the fusee arbor down to its seating, the old brass pins having been cut off level. While broaching the ratchet, hold it in a duster in the fingers only. Mark and drill two new pin holes through the ratchet deep into the fusee body. File up and drive in two brass pins. Cut them off level. Then put the fusee by its top pivot in a split chuck in the lathe and turn the pins off flush. Also turn out the centre of the ratchet like the old one, cutting out a complete ring around the arbor. This space is

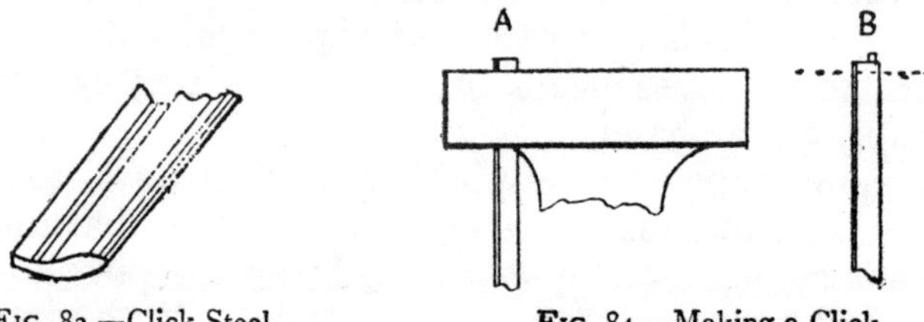

FIG. 83.—Click Steel.

FIG. 84.—Making a Click.

to accommodate the centre boss of the maintaining ratchet. If the new ratchet is so thick as to stand up above the edge of the fusee body, thin it by turning while in the chuck.

In a pair of turns, to centre a fusee properly, a "fusee arbor" is required. This is an adjustable holder in which the fusee square is held and admits of being centred truly by adjusting screws.

A broken fusee click must be first removed by placing the maintaining rachet over a hole in a stake and punching it through from underneath with a small flat-ended punch. New clicks are made from "click steel." This is steel rod, drawn to the section of a click, like Fig. 83. To make a click, screw the length upright in the vice, leaving a small piece standing up above the jaws, as at A (Fig. 84). Resting the safe edge of a pillar file on the vice jaws, file round it, leaving the pivot as at B. File the pivot up, without removing from the vice, until it goes nicely and easily into the hole in the maintaining ratchet. Then remove the length of steel, and with a slitting file saw off the click, as marked at C. Hold it in a pair of cutting nippers and flat the rough sawn surface with a file. Then place it in position and rest it on boxwood, click down, and file off the projecting part of the pivot. Transfer it to a flat steel stake, and with a round-faced punch slightly rivet its end into the countersink of the hole, taking care to only rivet sufficiently to prevent it coming out and not enough to tighten it. It must be perfectly free. If riveted a little too tight, it must be placed over a larger hole in a stake so that no part of the click is supported, then a tap given to the under-side of the riveted pivot will ease it; too heavy a tap will drive it through. When correct, lay it, click upwards, on boxwood or cork and file the top down level with the other clicks and the springs. In doing so carefully avoid filing anything but the click. Finally, cut the file burr off all round the edge with a sharp graver.

Fusee stopwork sometimes is troublesome. The stop A, Fig. 85, sometimes sticks up close to the plate, and will then stop the winding at every turn of the fusee. It should be perfectly free in its joint, and the pin must fit tight in the brass joint and loose in the stop. The slender spring should be in its place and carry the stop away from the plate promptly. This stopwork is designed so that when the chain reaches the top and last turn of the fusee it raises the stop A to the plate and the projecting finger on the fusee comes into contact with its end. The end of A should be *flat*, not notched or sloped, or the fusee finger may slip off it. If the watch overwinds, first ascertain if the fusee finger passes *over* or *under* A. This done, if it passes over A, bend A up a trifle so that it rises more quickly, if under, bend A down. Brass-nosed pliers should be used for bending.

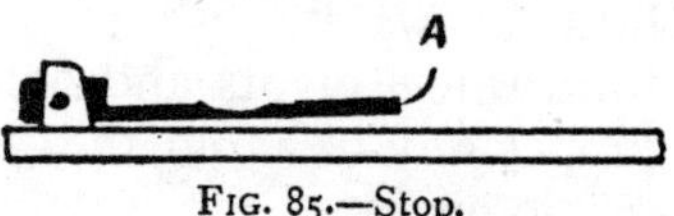

FIG. 85.—Stop.

The maintaining detent often fails to properly catch the teeth of its ratchet. If so, file its point up nice and sharp. If it slips up over the teeth it can be bent down with brass-nosed pliers while in the plates. If the fusee is pinned up too tight the maintaining work will not act. Copper wire should always be used for pinning up a fusee, as it is soft. If too tight, hold the main wheel in the hand and hit the bottom fusee pivot with a hammer, hard. This will slightly bend the copper pin and ease the fusee work. The interior of a fusee should be well oiled.

Safety Pinions, etc.—All English watches of a few years ago had fusees, and consequently wound to the left. In fact, this winding to the left got to be the distinguishing mark of an English lever with the general public. If a watch wound to the right it was at once put down as foreign. Some makers of English going-barrel watches, therefore, go to the pains of making them wind to the left, and to attain this end—merely to pander to a popular error—introduce a large dummy wheel, or idle wheel, between the going barrel and the centre pinion. Others introduce a pair of steel winding wheels under the dial, which are generally the cause of very rough and bad winding. Of the two the dummy wheel is to be preferred, as it only wastes a little power. But a plain going barrel, winding to the right, is better than either. It is claimed that the dummy wheel acts as a buffer in case of a broken mainspring, and takes off the shock from the centre pinion. If it does so at all it is very little, as watches so made are frequently found with broken third wheel pivots and bent teeth after such a breakage. The proper way to avoid these injuries is to use a safety pinion.

Some American watches have a safety barrel instead, which acts just as well. In them the barrel is of steel and is separate from the main wheel. In winding the watch the steel barrel is turned round. If a spring breaks the barrel receives the shock, and, being separate from the main wheel, the train escapes. These steel safety barrels are a little awkward to put hooks in. Very good fitting steel hooks must be fitted, and left as short inside the barrel as possible.

Clickwork.—Some English going-barrel key-wind watches have very bad winding work. The click is often only held by a small chamfer-headed screw (Fig. 86, A) which, when screwed home, holds the click tight so that it sticks, and to allow the click to work easily, must be left loose. This loose screw is in constant danger of working out. The remedy is to broach out the hole in the click parallel and then

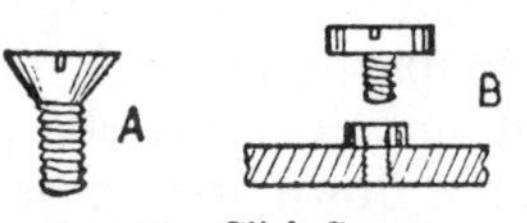

Fig. 86.—Click Screw.

screw a fixed brass stud into the plate for it to work upon. The stud can be drilled in its centre and a flat-headed screw fitted to hold the click down, as at B, or a small brass covering cock may be made instead.

A new steel winding ratchet or a new click to these watches should always be hardened and tempered.

Bar Geneva watches have "side clicks;" that is, click and click spring made in one and screwed to the side of the bar, as in Fig. 1, p. 1. When these fail to hold, file up the ratchet teeth, point up the click, and let it into the bar deeper by filing away the brass a little where required. A ratchet that has many broken teeth or that is badly worn necessitates a new barrel arbor to make a good job. But a makeshift can be arranged by turning off the ratchet all but a thin flange to hold in the bar, and fitting a loose new ratchet, carefully opened out in its centre hole to drive tight on to the square.

Three-quarter plate key-wind Geneva watches usually have circular click springs, as in Fig. 3, p. 3. When one is not readily obtainable of the correct size, a brass one can be easily made, and, if well hammered, is as good as steel.

CHAPTER IX.

DEPTHS, TRAIN WHEELS, ETC.

Depths.—The power of the mainspring is transmitted to the escapement through a train of wheels and pinions. To transmit the power evenly and with as little friction as possible the teeth are accurately cut to special curves. The care bestowed by the makers upon the form of the wheel teeth would be thrown away if the wheels did not engage with the pinions to the right depth. A wheel and pinion action is termed a "depth." When they are too close together the depth is said to be "too deep;" when too far apart it is "too shallow." In faulty watches there are thus deep depths and shallow depths. A deep depth causes waste of power and binding, the teeth of the wheel having little or no shake or freedom between the pinion leaves. A shallow depth causes friction and rapid wear of the teeth and pinion leaves. In a shallow depth there is too much shake between the teeth.

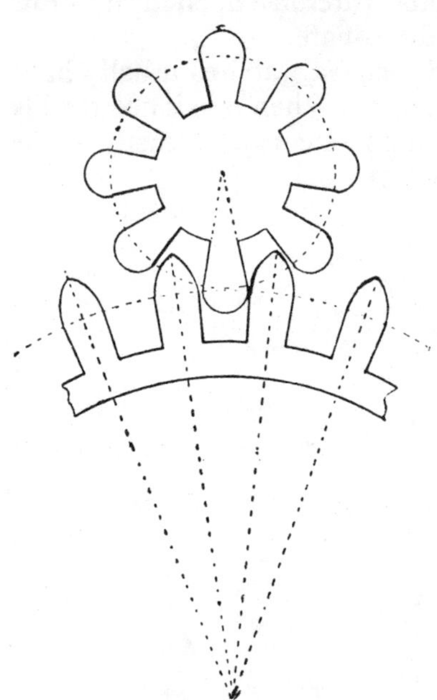

Fig. 87.—Wheel and Pinion Depth.

Fig. 87 shows roughly the construction of wheel teeth and pinion leaves. The acting diameter of the wheel is its "pitch diameter," and is shown by a dotted circle called the "pitch circle." Up to this point the flanks of the teeth are radial lines from the wheel centre. Beyond this the teeth are curved to a point. The flanks of the pinion leaves are also radial lines up to its pitch circle, and beyond that are finished off with a semicircle. In a correct depth the pitch circles of wheel and pinion are in contact with one another, as in the figure. The broadest part of the wheel tooth acts upon the broadest part of the pinion leaf. In this position the wheel teeth will have a little shake between the pinion leaves.

Some of the depths in watches can be easily seen, such as the barrel and centre pinion, or fusee and centre pinion, or the centre and third pinion of an English watch. Others can only be tried by putting the two wheels together between the plates. The pinion is held firm while the shake of the wheel teeth is observed, or felt. Some watches have sight holes drilled in the plates to enable all the depths to be seen; but in most some of the depths can only be felt.

A depth that is so deep as to have no shake at all will stop the watch, and must be corrected. A very shallow depth will not often stop a watch, but wastes the power and renders it liable to stop from the slightest cause.

When the pivots run in brass holes a depth can be corrected by drawing one pivot hole, that next to the pinion being the best. A hole is drawn by broaching it out larger and pressing the broach to one side firmly with the finger-tip as it is turned, causing it to cut on one side faster than the other, and draw the hole oval in the desired direction, as in Fig. 88. It is then broached out round and "bushed" (see p. 90).

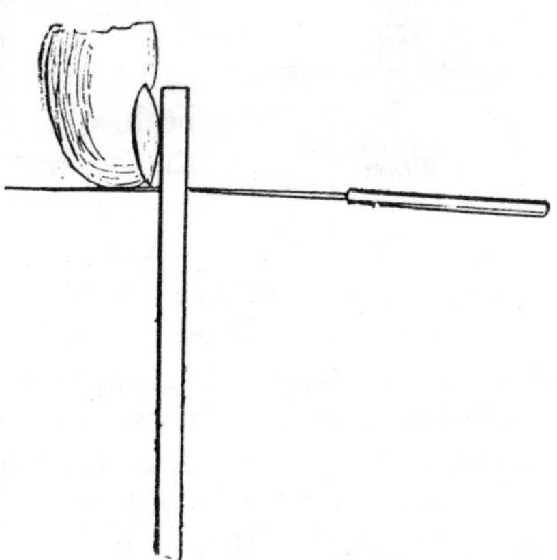

FIG. 88.—Drawing a Pivot Hole.

When the pivot holes are jewelled, a deep depth can be corrected by topping the wheel teeth in a "wheel rounding-up tool." This is a kind of automatic wheel-cutting engine, and extremely useful to a watch repairer. In effect it makes the wheel a little smaller. Topping a wheel with a file is no good at all. A shallow depth, when the holes are jewelled, can be sometimes improved by stretching the wheel rim by gentle hammering or burnishing; but this makes a poor job. The only real remedy is a new wheel.

With constant wear, pinion leaves get cut by the wheel teeth into holes or hollows. These hollows hold the dirt and stop the watch, besides causing great friction while the watch goes. It might be thought that when two metals work together, such as brass and steel, as in a brass wheel and a steel pinion, or a steel pivot in a brass hole, that the softer of the two would wear the most. It is the contrary that always happens. The harder metal wears more quickly. This is because wear is caused by dirt, dust, or grit getting between the surfaces. This imbeds itself in the softer metal, and converts it into a kind of grinding-mill, which cuts the harder metal. Thus pinion leaves get worn out before the wheel teeth, steel pivots wear before the brass holes, and brass scape-wheel teeth cut hard steel pallets. It follows that watches wear much longer if kept clean.

A worn pinion cannot be repaired. Sometimes, where there is room in the watch, the wheel which runs with it can be raised or lowered so as to act on a fresh and unworn part of the pinion, as in Fig. 89, which shows a section of such a wheel. The doming stake (see p. 76) can be used. The wheel is placed in the hollow, with the pinion passing through the central hole, and a punch is placed over it, having a central hole to accommodate the arbor. A tap or two will spring the wheel concave. This process sometimes leaves the wheel true and sometimes not. If not, it must be run in a pair of calipers, or in the turns, and the high or low spot noted. Then lay the wheel on a piece of boxwood, and a tap with a small flat punch on the centre of one arm will *raise the wheel rim* at that point. A few taps and trials will soon get it true. Any untrue wheel can be treated in this manner.

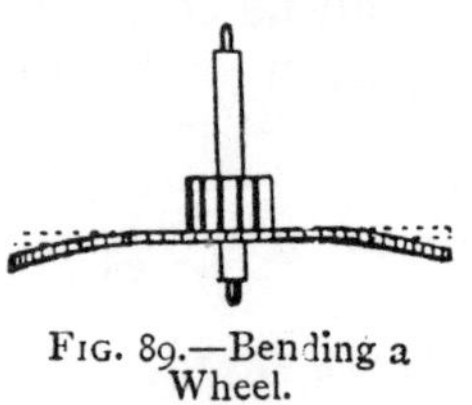

Fig. 89.—Bending a Wheel.

Pivots.—Pivots often get worn, ribbed, and cut into channels like Fig. 90. These must be polished down in turns or lathe, and the holes which then would be too large must be bushed.

To smooth and polish a pivot, put it in a split chuck in the lathe, holding it by the other end of the arbor, or, if there is no arbor to grip, hold the wheel in a step chuck. If neither of these methods can be used, the wheel must be cemented by shellac to a small brass chuck. Warm the chuck in the spirit lamp and run shellac upon it. The chuck face should be flat with a central hole. Lay the wheel upon the chuck face and again warm until the shellac runs. When partially set, screw the chuck in the lathe. Now hold the lamp so that the flame warms the chuck and softens the shellac, and run the lathe slowly. When softened, hold a peg point to the arbor close up to the pinion face or wheel seating, in the angle. Touch it very lightly and the peg point will run it true. Continue to revolve the lathe and apply the peg point until the shellac sets. Fig. 91 shows what is meant. It is possible to hold nearly

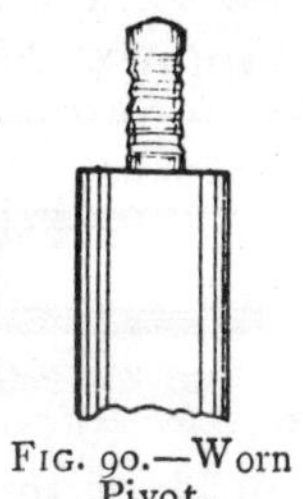

FIG. 90.—Worn Pivot.

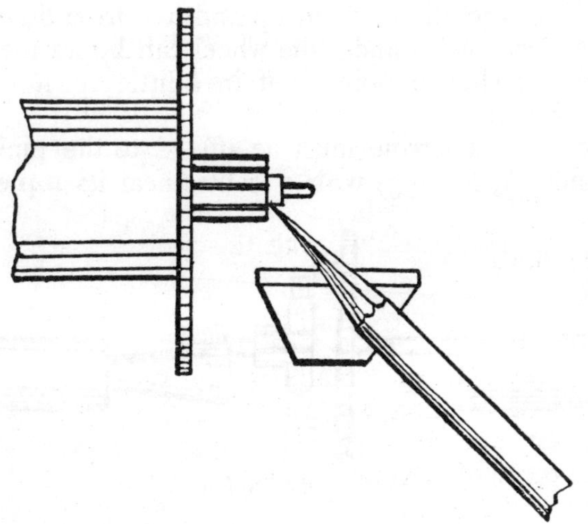

FIG. 91.—Running a Wheel True.

every job in this way in a watch lathe, and be independent of chucks. Things held thus run perfectly true, which is not always the case with split chucks. The whole process of cementing and running true only occupies a minute, though it takes many to describe.

When true in the lathe, take a strip of flat steel about 6 in. long and $\frac{1}{8}$ in. wide; file it flat on one side and slightly bevel one edge to the section shown in Fig. 92. This is a polisher. Mix up some oilstone dust and oil to a paste on a polishing stake. Apply some to the polisher face, and while the lathe is running polish or grind the pivot down, using the polisher as a file with light pressure, as in Fig. 92. Care must be taken to hold the polisher level, or the pivot will be ground down tapered instead of straight. A very little of this treatment will smooth the pivot. Too much will reduce it to nothing. When smooth, clean off the oilstone dust from the pivot with pith and cleanse the polisher, giving it a fresh surface with a smooth pillar file. Recharge it with red-stuff and oil mixed to a paste on another polishing stake. Use in the same manner for a few minutes, revolving the lathe fairly fast. This will polish it. Keep the mixed and unmixed red-stuff very carefully covered up and free from dust.

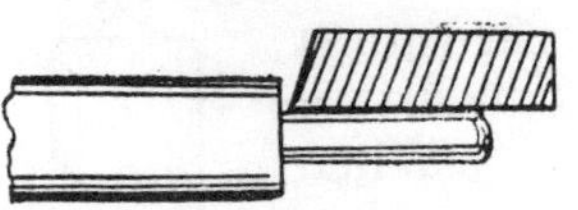

FIG. 92.—Polishing a Pivot.

If shellaced on a chuck, the wheel can be warmed off and the remaining shellac boiled out in a little spirit in a spoon held over the lamp.

In the turns, a ferrule must be affixed to the pinion, and a back runner (A, Fig. 93) with a centre near its top edge used.

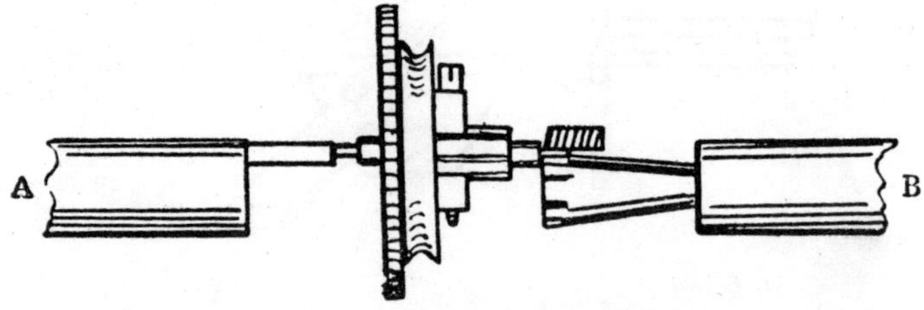

FIG. 93.—Polishing a Pivot in the Turns.

A polishing bed (B) is a brass runner turned to the shape shown and having slight hollows cut around its edge for various sized pivots to lie in. Select a hollow that takes half the pivot in depth. Adjust the runners and pinion as shown in the figure, and use the polisher as before described. On the downward stroke of the bow the polisher must go forward, on the up stroke of the bow the polisher comes back as in pin filing. Never forget to apply a little oil to the back centre. After polishing in this manner a few minutes, reverse the motion for a few strokes, letting the polisher come forward as the bow comes down. This mixes up the polishing paste.

Oilstone dust should be used rather thin and "sloppy." Red-stuff should be rather a stiff paste, and a very little put on the polisher. The art of producing a good polish on a pivot is to move the polisher about upon the pivot, not keeping it always up to the shoulder, to use long strokes of the bow (or in a lathe rapid revolution) and long strokes of the polisher.

A final gloss can be put on a pivot by cutting a peg flat on one side and applying red-stuff, using that as a polisher for a moment with a high speed.

Turning Pinions.—When a new pinion is required, a rough one (Fig. 94) is purchased from the tool shop. This is hardened and tempered ready, and the leaves are cut and polished. It is also centred truly ready for turning. First put the old pinion in lathe or turns, turn off the riveted face of the pinion, and knock the wheel off by punching the pinion through it. Turn the new pinion body (leaf portion) down to the correct length by using a good speed, a sharp graver, and cutting with the point only. The ends of the leaves can be faced well enough for most jobs by a polisher held as in Fig. 95, using first oilstone

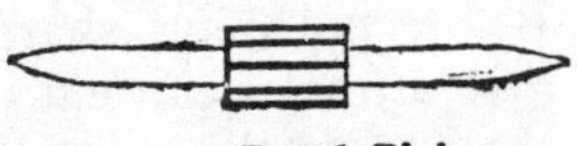

Fig. 94.—Rough Pinion.

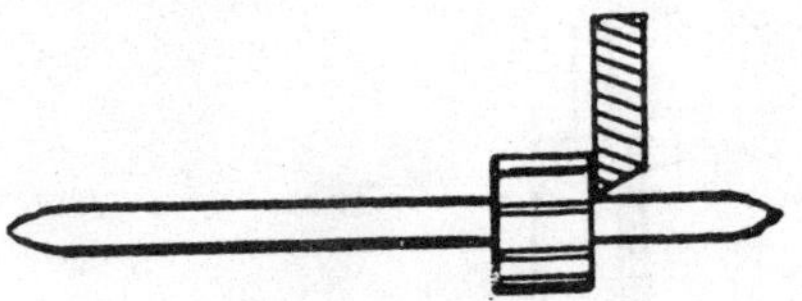

Fig. 95.—Facing Pinion Leaves.

dust and then red-stuff as before described. If a perfect flat face is required, as in new work, the centre of the leaves must be turned hollow or "cupped" a little, to leave no root in the shoulder, and a square or bevel edged polishing lap brought up to it in a watch lathe. A polishing accessory is made to fit in the hand rest of the lathe, by means of which these laps can be accurately adjusted. A seat for the wheel is then turned so that the wheel drives on tight. It must on no account be an easy fit. The points of the leaves should stand up above the surface of the wheel for riveting over and be turned hollow inside, leaving a circle of points like a crown, as in Fig. 96. The pinion arbor should be turned down, leaving a slight square shoulder, as at A, Fig. 97, for the polisher to go against when polishing it. The arbor should be finished and polished before turning the pivots. A straight pivot should have a clean square shoulder, as at A, Fig. 98; not a "root," as at B. Turn the pivot as smooth and true as possible, then smooth it with oilstone dust. Before polishing, sharpen the graver and place as in Fig. 99, to give a final clean cut in the shoulder—just one sweep down of the bow in the turns or two or three revolutions of a lathe.

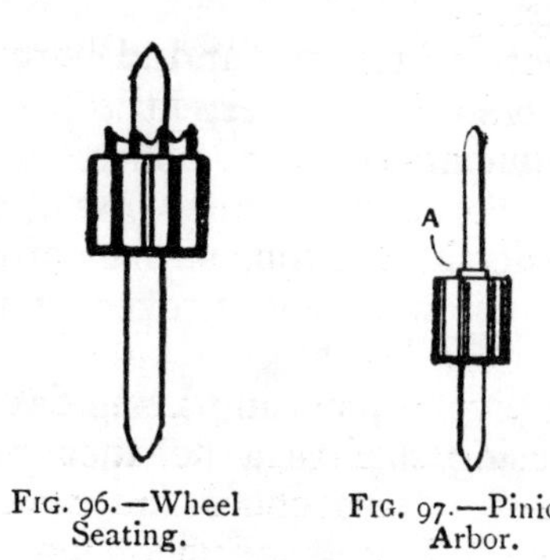

FIG. 96.—Wheel Seating.

FIG. 97.—Pinion Arbor.

Turn the bottom pivot first, and, if a jewel hole, fit it to the hole by reducing with oilstone dust until it just goes in and sticks. Then polish, and it will fit properly. If a brass hole,

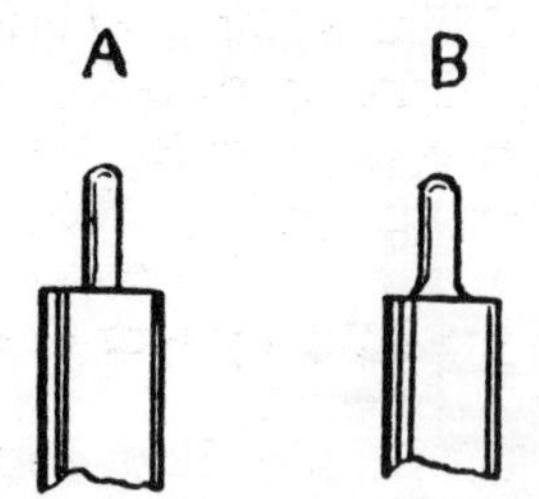

FIG. 98.—Square Pivot Shoulder.

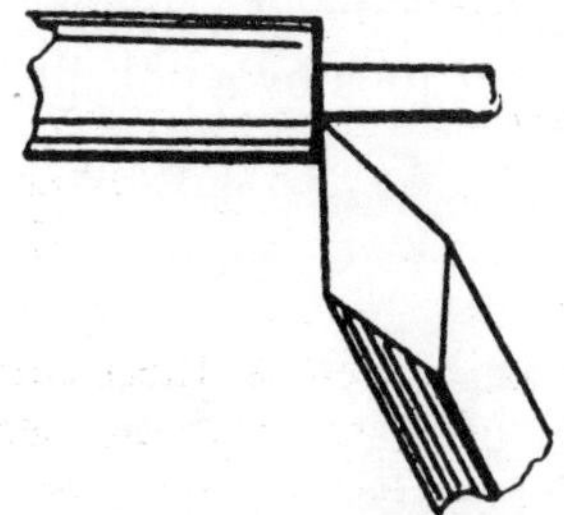

FIG. 99.—Cutting the Shoulder.

leave the pivot a shade large and open the hole to fit. When the bottom pivot is turned and the wheel and pinion got to the correct height, the total length of the pinion can be measured and the top pivot turned. Take a piece of thick brass wire, file it flat on both ends, and fit it accurately so that it will stand up on the bottom plate or jewel hole and have a correct end-shake under the top plate or hole. This will then be a gauge for the total length of the pinion, and can be measured over all with vernier gauge or douzième gauge.

Round up the ends of the pivots by a pivot file and burnish them. Finally rivet the wheel on. Put the pinion in a steel stake in the vice jaws. Press or punch the wheel on to its seating, and with a flat-ended steel punch rivet the points of the leaves well into the wheel. As a finish, turn the riveted surface clean.

Using a lathe, the pinion body can be held in a split chuck for reducing and facing, and also for turning the wheel seating, turning and polishing the arbor and the pivots. Or it may be shellaced into a chuck like Fig. 100. This chuck has a cone hollow, and is filled with shellac. This is warmed and the pinion pressed in until its lower end rests in the bottom of the

FIG. 100.—Cone Cement Chuck.

cone. The shellac is built up round it and the pinion "run" true with a peg, as shown in Fig. 91, p. 85.

A cone shellac chuck for "running" true has to be warmed as it revolves in the lathe, and the ordinary form of chuck carries the heat away very much to the lathe headstock. This difficulty can be got over by drilling two large holes through the chuck at right angles to and through each other, and broaching them out as shown in Fig. 100. The chuck end that is heated is then only joined to the lathe by four slender strips of brass, and the heat is retained better.

These cone chucks ensure the perfect truth of both ends of the pinion.

If preferred, the entire pinion can be turned between dead centres, using a loose pulley runner and a carrier, as in Figs. 49 and 50, p. 44.

In the turns, a ferrule is affixed and the pinion placed between large centres to turn the body, face it, and turn the wheel seating. The arbor is turned and a rough pivot formed on its end, to rest in the polishing bed while polishing the arbor, as in Fig. 93, p. 86. It is cut to length by turning through with light cuts and a very sharp graver, like Fig. 101. This

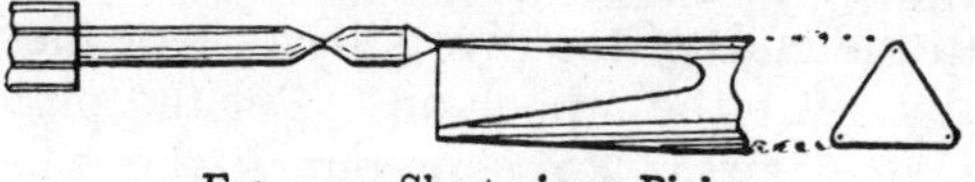

FIG. 101.—Shortening a Pinion.

will leave almost a point. The centre is "run" by a pivot file, as in Fig. 102, A showing the end of the "running" arbor. This forms a true turning centre. For turning the pivots a

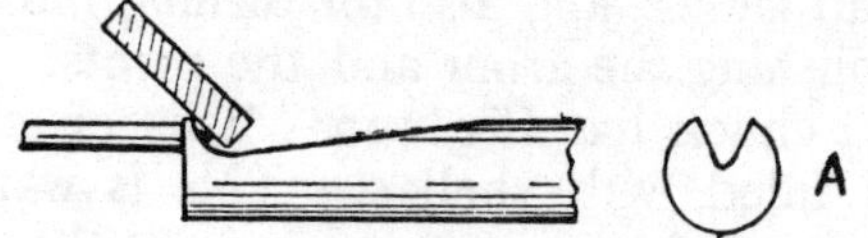

FIG. 102.—Running a Centre.

back centre (A, Fig. 93) is used. The front runner is triangular with a minute centre (made with pointed chamfering tool) at each corner, as near the edge as possible, like Fig. 101. They are polished as shown in Fig. 93.

Bushing Holes.—To bush a pivot hole, "bouchons" like Fig. 103 are used. These are small lengths of hard brass wire, each end of which is turned down and drilled up its centre. Select one with a central hole that will just stick on the pivot. Hold it in a pin vice and taper it very slightly at the point. Open the pivot hole by broaching from the inside surface of the plate or bar, until the bouchon goes in tight half-way. Then insert it and break it off where half cut through (A, Fig. 103). File the projecting end flat, place a flat punch upon it and tap it quite in. Open it with a broach to take the pivot nicely, taking care to broach the hole upright. The inside surface may be stoned off level, or it may be turned in the mandrel. The outside surface where the oil cup comes may be chamfered out or turned. In each case turning makes the best job.

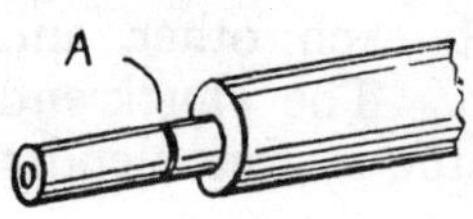

FIG. 103.—Bouchon.

The bottom centre wheel hole in a full-plate watch, or either centre hole in a Geneva, must have a specially turned bush. A piece of clock bushing wire is sawn off about twice as long as the watch plate is thick. It is held in a pair of hand fitting tongs (Fig. 104) and broached out nearly to go on the pivot, placed on a turning arbor in the turns, and made true at each end, reduced in diameter and tapered. The pivot hole is then broached out to receive it. The bush must be turned to the exact length to rivet in, each face being turned a little hollow to leave an edge for riveting over. The hole is very slightly chamfered, and the bush riveted in with a flat punch on a flat stake. It is finally turned off on both sides flush and clean, and opened out to fit the pivot.

The bottom centre pivot of a full-plate watch is often so worn that turning it true and polishing will reduce it to a smaller diameter than the part where the cannon pinion fits on. In such a case turn the pivot down level. Make a steel collar and soft solder it on, afterwards forming a new pivot. The hole

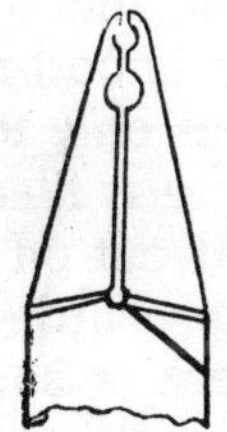

Fig. 104.—Hand Tongs.

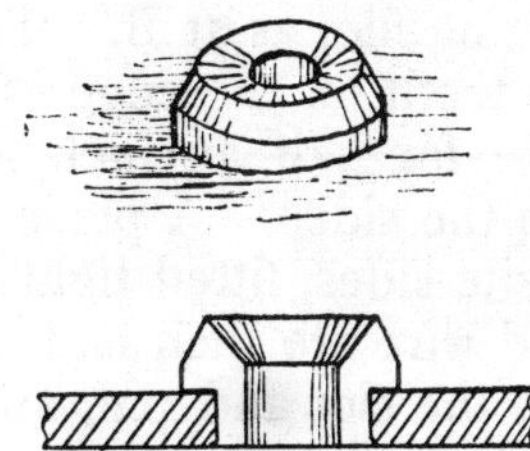

Fig. 105.—Top Fusee Bush.

can then be opened out larger to fit. The steel collar is conveniently made by turning down a portion of an old cannon pinion, and is then ready hardened and tempered. The collar forms the new pivot.

Similarly, a top fusee pivot is often worn smaller than the winding square. In this case the pivot must be turned down and polished and the square reduced, the set-hands square being also reduced to match it and fit the same key.

A "hollow" fusee is one in which the pivot is sunk in a hollow and cannot be got at for turning without taking the fusee to pieces. In these fusees the body can be entirely removed by taking out the three screws that hold the internal ratchet on.

A top fusee hole in a full plate requires a bush specially turned from bushing wire as a centre wheel. Turn it to the shape shown in Fig. 105. The bush rests upon the shoulder.

and is riveted on the under side as shown in the section. A "setting" held by three screws, as in many ¾-plate watches, may be bushed as described for a centre wheel.

Any bent wheels should be trued by a punch on boxwood, as on p. 84. If they touch their sinks in the watch plate, put the plate in the mandrel and turn the sinks out.

Wheel Teeth.—Broken teeth are put in by slotting and soft solder. The root is filed out flat, as A, Fig. 106, slotted

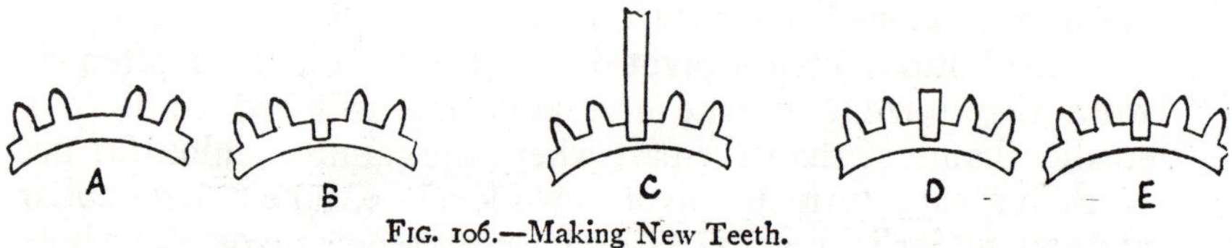

Fig. 106.—Making New Teeth.

with a slitting file, as at B. To widen the slot and square its sides, the teeth on the edge of the file can be smoothed off on an oilstone for half an inch at one end, so as only to let the file cut on the sides. A piece of brass pin wire is filed flat on two opposite sides, fitted tight in the slot, and cut off as at C. Wheel and wire are then laid on a blueing slip, a little solution applied to the slot and pin, and a minute piece of solder laid on it. All is then warmed over the flame until the solder runs in. After well washing, file off as at D, clear out the solder between the teeth, and shape up as at E. Flat both sides by laying the wheel on cork and using a fine sharp file. Finish with a $\frac{3}{0}$ buff and a rouge buff.

Fig. 107.—Three New Teeth.

Several teeth can be put in a wheel by filing out and fitting a piece, as in Fig. 107, but in such cases a new wheel is best.

New Pivots.—Broken pivots are replaced by drilling, fitting a piece of tempered steel like a plug, and forming a new pivot.

A top centre pivot, like Fig. 108, A, is best put in by turning off the arbor and drilling the pinion body itself with a large hole, forming arbor and pivot on the new piece of steel, as at B. A third pivot is replaced as at C. A shallow pinion and thin arbor like D is repaired by drilling right through and putting new arbor and both pivots. A broken pallet staff pivot in an English watch requires a new staff complete.

In a lathe nothing is simpler than putting in pivots. Where possible the pinion body (preferred) or arbor is held in a split

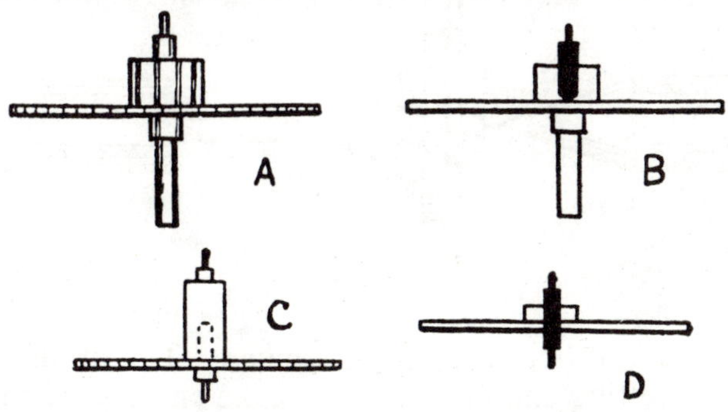

Fig. 108.—New Pivots.

chuck; or the wheel is held in a step chuck or by shellac. The hand rest is brought up and a centre for drilling *turned*, like Fig. 109. A very sharp graver and a good light are necessary, to see that no "pip" is left in the bottom of the hole. A drill (hardened and not tempered) well sharpened is put in the drill holder and lubricated with turps. Slow revolution and good pressure will cut almost any pinion. The

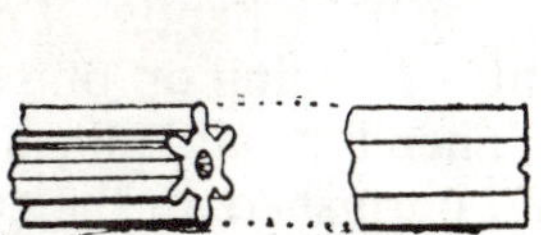
Fig. 109.—Drilling Centre.

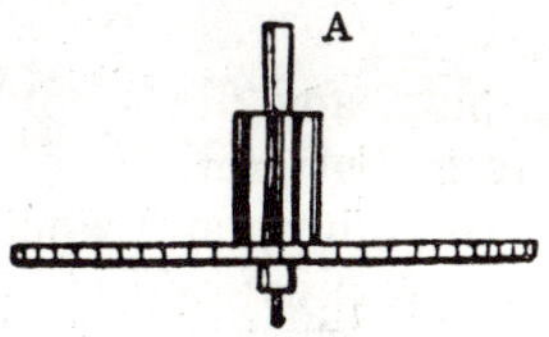

Fig. 110.—Making New Pivot.

moment the drill stops cutting, resharpen it. If it cannot be made to cut, blue the pinion in the spirit lamp and remove the blue by dipping for a moment in strong spirits of salt and well washing. It will then drill all right. Drill at least two or three diameters deep. Then peg the hole clean. Take a needle, let down to a blue temper, and in the pin vice file it down tapered (nearly straight) to fit the hole tightly half in. Fit it in. Cut it off and file flat on the end, as at A, Fig. 110, and hammer it in. Then put again in the lathe and turn, fit it to the hole and polish, finally burnishing the end.

In the turns there is more trouble. Fit up the turns as in Fig. 111. The back pivot runs in a centre, and the drill is

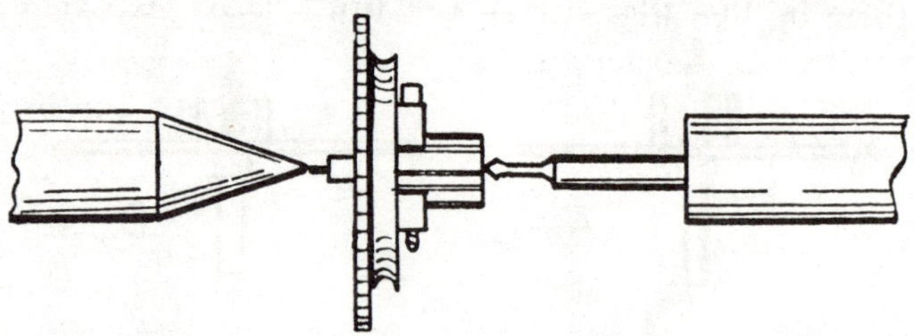

FIG. 111.—Drilling a Pinion.

held in a central hole in a brass runner. The drilling centre is made by resting the pinion on a stake and chamfering with a pointed tool in its exact centre. If not quite correct it can be drawn and chamfered deeper. Then drill, leaving the brass runner loose and keeping the pressure up by hand. Fit the

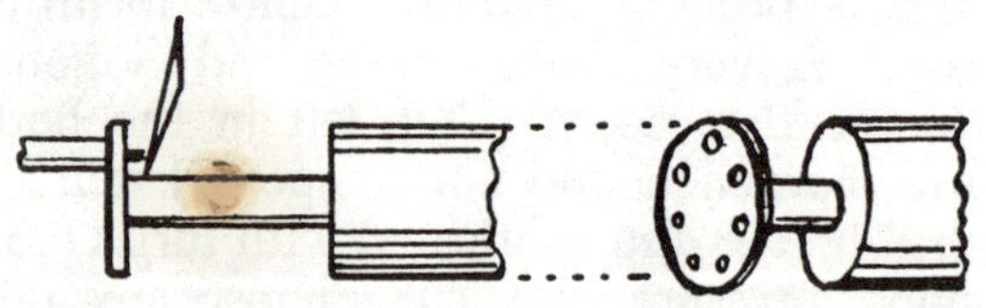

FIG. 112.—Lantern Runner.

steel plug, and file a turning centre upon it by hand. Try in the turns for truth, and draw the centre by filing on one side until the pinion and wheel run true. Then turn and polish the pivot. A pivot end is rounded up and burnished in the turns by means of a lantern runner, as in Fig. 112. These are made

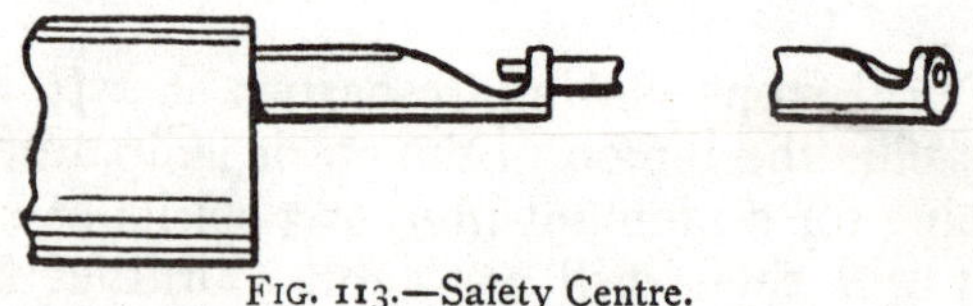

FIG. 113.—Safety Centre.

of brass. A slender back pivot may be put through a "safety back centre," like Fig. 113; here the pivot shoulder takes the thrust. A centre like this is always used with a coned pivot.

Jewel Holes.—Jewel holes often crack. A cracked hole will cut a pivot and cause friction; it should be replaced with a new one. Some jewel holes are set in the plate, lying in small circular sinks, and a thin brass edge is burnished over to hold them in, like Fig. 114, at A. When such a hole is broken it can be pushed out with a flat-ended peg and a hammer. This generally raises the thin brass edge; but if not, it can be raised by running a centre-punch point round inside with a circular motion. Then select an unset jewel hole to fit the sink and pivot, and burnish the brass edge over it again with the centre-punch point held in the fingers, working it lightly round and round, as in Fig. 115.

A jewel hole that is fixed in a brass setting and held by two screws had better be fitted by a watch jeweller. Watch

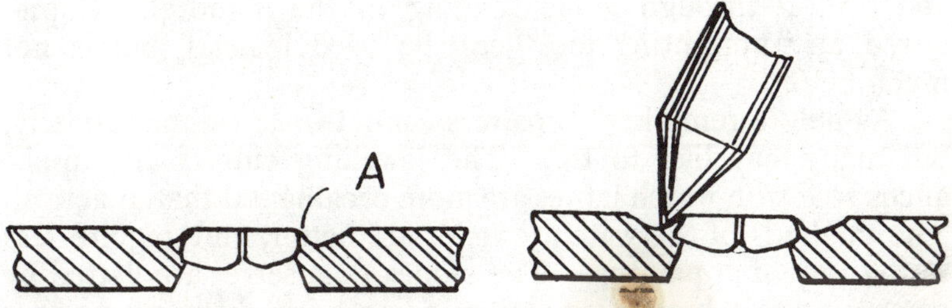

Fig. 114.—Jewel Hole. Fig. 115.—Burnishing a Setting.

jewelling is a trade by itself, and very few ordinary repairers can do a jewelling job even passably; none can do it well.

English jewelling fitted with endstones can only be replaced by keeping a stock of set jewel holes and selecting one that fits the sink and the pivot. Even then the thickness is often wrong. This jewelling also is best sent to a jeweller for repair. An endstone with a crack across or a centre "pip" cut by the pivot end must be replaced.

A Geneva endstone set in a small brass slip, when broken, can be replaced by fitting a loose one. Chamfer out the old setting with a circular chamfer and select a loose endstone that lies in the hollow with its surface level with the brass slip. A similar method can be followed in the case of an English set endstone, but is not so good as a new setting and stone.

An English endstone slip, fitting in a dovetail groove in a potance, is replaced by filing up and fitting a new brass slip.

Such slips are purchased in the rough with stones set in them. First flat the surface of the slip level with the endstone, then bevel and taper its sides, reducing the width gradually until the slip goes in its groove tight with the endstone central. Then cut off and trim the ends.

A broken centre wheel jewel hole or fusee hole is expensive to replace, and often is best broached out and bushed with brass. Occasionally, also, in a cheap watch a third or fourth jewel hole may be filled up with brass and re-drilled. In this case, fill up with brass, rivet well in, and smooth off by stoning or turning level. Then put the frame together, centre it in the mandrel by the other hole, and mark the hole to be drilled by catching the centre of revolution with a sharp graver point. When marked, hold a pivot drill in a drillstock in the hand and drill it through while running in the mandrel. If preferred, an "uprighting tool" can be used instead, but is not much better.

As before remarked, repairers cannot do jewelling properly, but many may like to try. The jewelling cutters and appliances sold with watch lathes are more ornamental than practical. The methods of the watch jeweller are better, and require less in the way of apparatus. The stones themselves are flatted by cementing to a brass disc and grinding with diamond powder on a copper lap running in the lathe. When one side is flat the stone or stones are reversed, and the other sides flatted and reduced to the correct thickness.

They are drilled by centring them separately in the lathe with shellac and using a diamond turning tool to cut a central sink. Then they are perforated with a diamond drill kept moist, or by a steel wire and diamond powder. The interior of the hole is polished with the finest diamond powder on a copper wire with extremely rapid motion. A jeweller's lathe has a foot wheel about 30 in. diameter, and a mandrel pulley of only 1 in. or 1½ in., driving with a cotton band—a mere thread—and revolving the foot wheel as fast as possible. The speed thus obtained is very high, being several hundred revolutions per second. The outside edge of the stone and the angles of the hole itself are turned out with a diamond tool.

To set holes in brass settings, jewellers screw a short length of brass rod in the lathe mandrel and drill its centre, turn out a sink to take the hole, turn the thin outer edge, insert the stone, and burnish it in; then cut the set hole off. It is then reversed and shellaced on the chuck and the other side of the setting turned. A jewel hole is accurately centred by warming the shellac as the lathe runs and running true with a peg point, as described on p. 85. To see if it is absolutely true, a peg point is inserted in the hole and rested on the hand rest. As the lathe revolves, any want of truth is shown by wobbling of the peg end, A. If true, A will be motionless (Fig. 116).

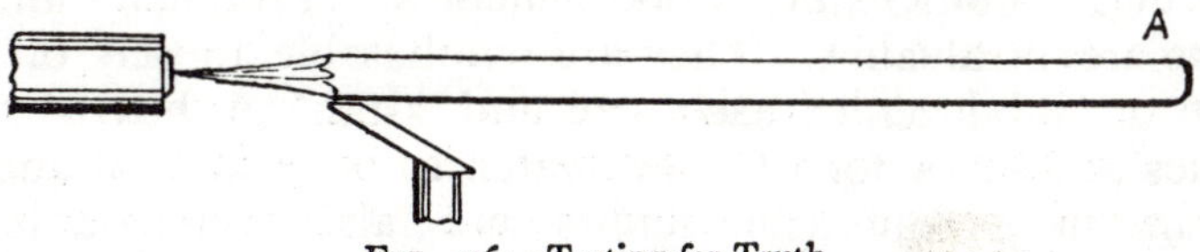

FIG. 116.—Testing for Truth.

The hand rest is used throughout, and cutters formed from the tail ends of old files like those shown in Fig. 117. These

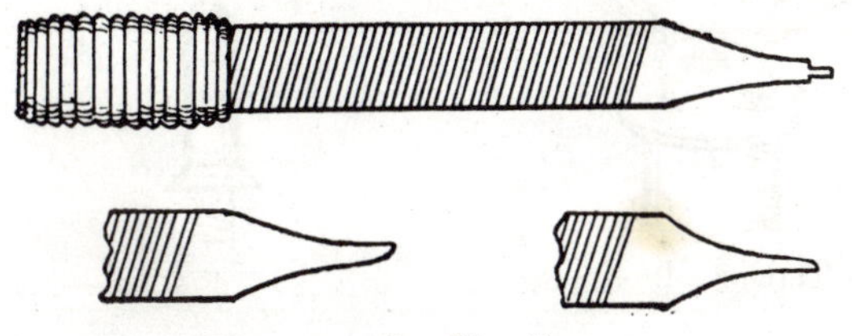

FIG. 117.—Jewelling Cutters.

cutters lie nice and flat on the hand rest, and are generally handy. Their handles may be wound with string and cemented with sealing wax.

All sinks for jewelling and screw heads are done in the same manner by shellacing on to a chuck and by these hand tools.

By using these methods a repairer may sink jewelling to decrease endshake, and may perhaps manage other details, but cannot expect to be able to do the entire process of watch jewelling.

Frames, Screws, etc.—In English full-plate watches the top plate is generally pinned on. Sometimes the pin holes burst through the pillars or the pillars split. It is then best to flat the pillar top and drill and tap it, fitting an ordinary pillar screw.

A cock or other screw, broken in its hole, is often difficult to remove. First try two graver points to turn the broken tap out, as in Fig. 118. Two points used thus relieve the screw of the friction against the sides of the hole. If nothing can be done in this way, it must be punched out from behind with a flat-ended hard steel punch that fits the hole. One hard blow, with the plate resting on a good stake, should drive it out.

The fitting of new screws is an important part of watch repairing. For Geneva work generally "Progress" finished screws are invaluable. They are purchasable in sets of half-dozen or one dozen of each size and kind. A box of these enables any screw for a Geneva watch to be picked out and put straight in. American screws are also purchasable for American watches, and save a great deal of time and trouble.

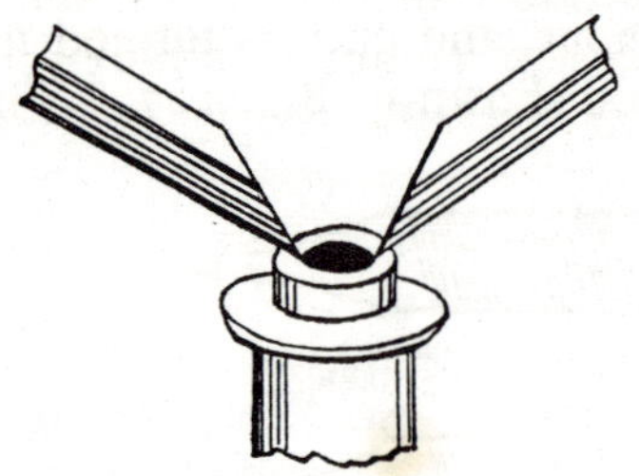

FIG. 118.—Turning a Broken Screw.

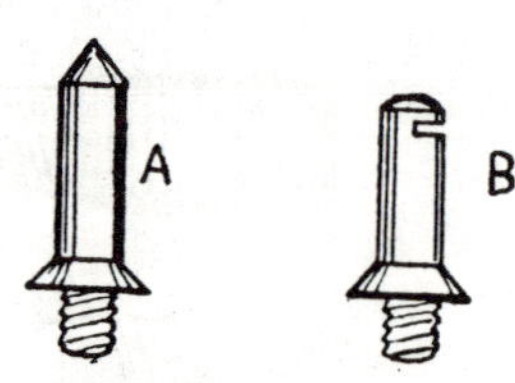

FIG. 119.—Cap Stud.

For English watches and some others of special make soft unfinished screws are bought. These require hardening and tempering, shortening to the correct length, and the heads reducing as required. But occasionally, with the best possible stock of screws, one will have to be made. This is easily done by putting a piece of steel wire in a split chuck, turning down the tap portion, and threading it in a screw plate. It is then reversed and held by the tapped portion while the head is turned and cut to length. Very small screws for hairspring studs, etc., and "shoulder" screws are the ones that most frequently have to be made thus.

Cap studs for English full-plate watches are bought in the rough, like Fig. 119, A, and require cutting to length and marking for the slot. This is then cut with a slitting file and the top end burnished in the lathe, as at B.

Screws are finished in three ways. They may be polished

bright, or they may be heated on a blueing slip to a "red" or a "blue." In fitting from "Progress" finished screws, it should be remembered that a "red" screw can always be blued if necessary by simply heating it.

To polish screw heads, hold them in a split chuck and rest a fine file on the roller T rest; follow this with a $\frac{3}{0}$ emery buff. Then smooth with oilstone dust and polish with red-stuff. For common jobbing a "burnished" head is good enough. After using the $\frac{3}{0}$ emery buff, apply a flat burnisher with good pressure and rapid revolution of the lathe. A burnisher is sharpened by rubbing it crossways with a fine emery buff.

For a bright finish this surface is good enough for jobbing, and does for screw heads, studs, winding and set-hand squares, etc. But a burnished surface is a false one, and will not blue. For blued or red screws the surface, for common jobbing, may be left from the $\frac{3}{0}$ buff. For better work the screws must be polished before blueing.

Loose pillars in a watch frame should be riveted tight by resting their tops on boxwood and using a round-faced punch. If the pillar tops are rested on a steel or brass stake they will be bruised, and the endshakes of all the wheels will be found to be closed up.

CHAPTER X.

ESCAPEMENTS.

Lever Escapement.—This escapement, the most important of all, is now found in nine out of every ten watches made. It may be in various forms, but all are made on one principle. Fig. 120 shows the form used in English watches; it is of the

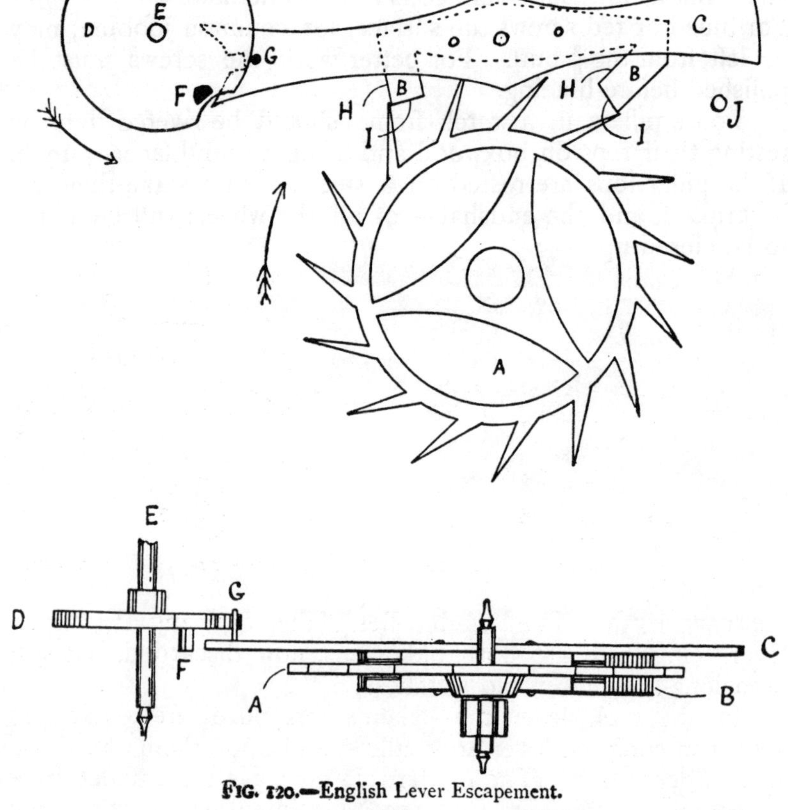

Fig. 120.—English Lever Escapement.

right-angled variety, and has pointed or "ratchet" teeth. Other forms are shown in less detail; the "straight line," in Fig. 121. In this the scape wheel, pallets, and balance are all planted in a straight line. Most Swiss and American watches are made thus. This figure also shows a "club-tooth" scape wheel, used in most foreign watches. The "pin-pallet" escapement (Fig. 122) has round pins for pallets, and the inclines are on

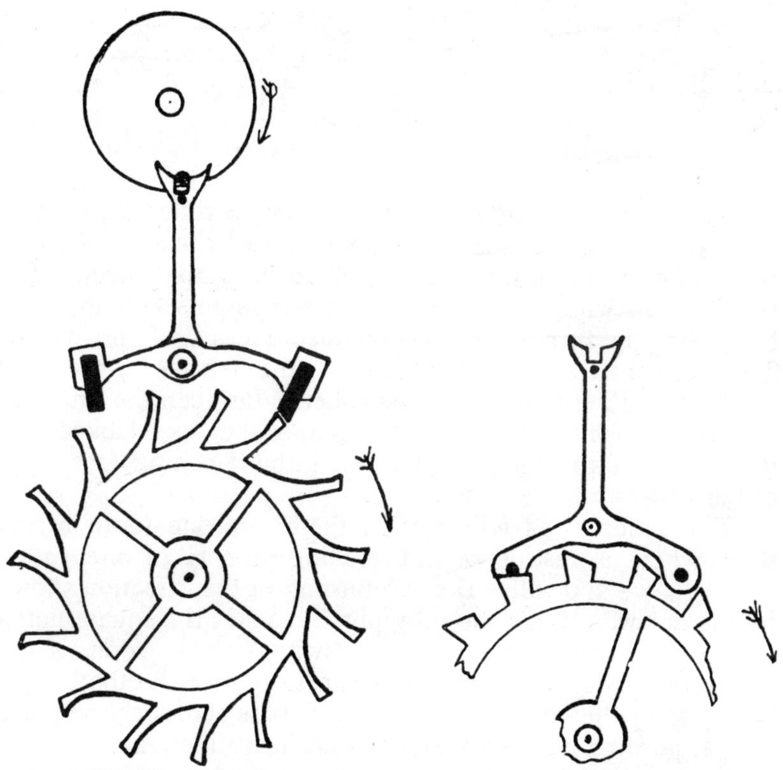

FIG. 121.—Straight Line Escapement. FIG. 122.—Pin-pallet Escapement.

the scape teeth. The "club roller" (Fig. 123) and the "rack lever" (Fig. 124) are both old forms, now discarded, but still occasionally found in old watches.

In the rack lever the balance was never free. It was always moving the lever and pallets, and upon them rested the scape-wheel teeth. The modern forms leave the balance free, except while unlocking and receiving impulse. This was the origin of the words "detached lever" engraved upon old watches. It was to distinguish them from the rack lever, which was not "detached."

Referring to Fig. 120, the lever escapement consists of a fifteen-toothed scape wheel, A; a pair of anchor-shaped

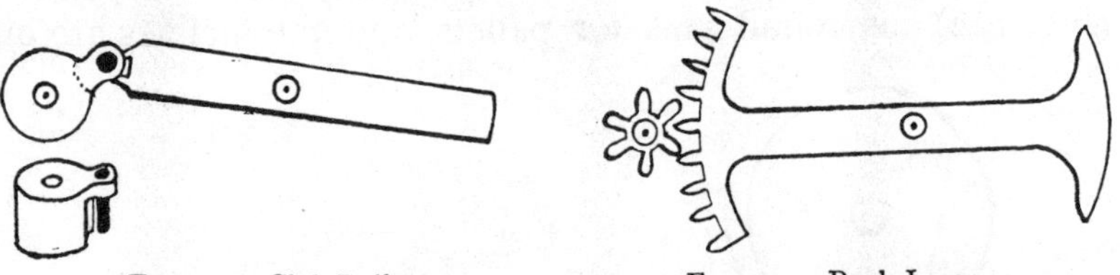

Fig. 123.—Club Roller.

Fig. 124.—Rack Lever.

pallets, B, fixed rigidly to a lever, C; and a roller, D, fixed to the axis of the balance E. The power of the mainspring is transmitted through the train wheels to the scape wheel. The teeth of this wheel are made to deliver impulses to the lever by passing in turn across the inclined faces I of the pallets. The lever passes these impulses on to the balance by means of a ruby pin, F, set upright in the roller. Between the impulses the balance spins freely, and the scape wheel is "locked" by the points of the teeth resting upon the "locking faces," H, of the pallets.

The action is as follows: In the figure one tooth of the scape wheel is resting on the locking face H of one pallet. The balance and roller D are returning in the direction shown by the arrow. When the ruby pin F reaches the square notch in the lever it will carry the lever with it. As soon as the lever and pallets have moved a short distance the tooth resting on the pallet will be released and pass across the impulse face I, pushing the lever across, and instead of the ruby pin leading the lever the lever will push the ruby pin, giving the balance an impulse. The tooth giving impulse finally drops off the pallet corner, and another tooth falls upon the locking face of the other pallet. The balance travels round three-quarters of a turn or so and then returns, going through the same series of movements as before, but in the reverse direction, and so on. The angular movement of the lever is limited by the "banking pins" JJ, which are placed in such a position as to hold the lever in the correct position for the ruby pin to enter the notch properly. The locking faces of the pallets are so shaped that the pressure of the scape-wheel teeth upon them tends to draw the lever against the banking pins and hold it there. There should be enough "draw" to keep the lever to the pins, but not enough to oppose any serious resistance to the balance in unlocking the teeth. In spite of the "draw," an accidental shake might move the lever and unlock the wheel. To prevent this there is a safety pin, or guard pin, G, set upright in the lever, which would come in contact with the roller edge. To allow this pin to pass while the lever is giving

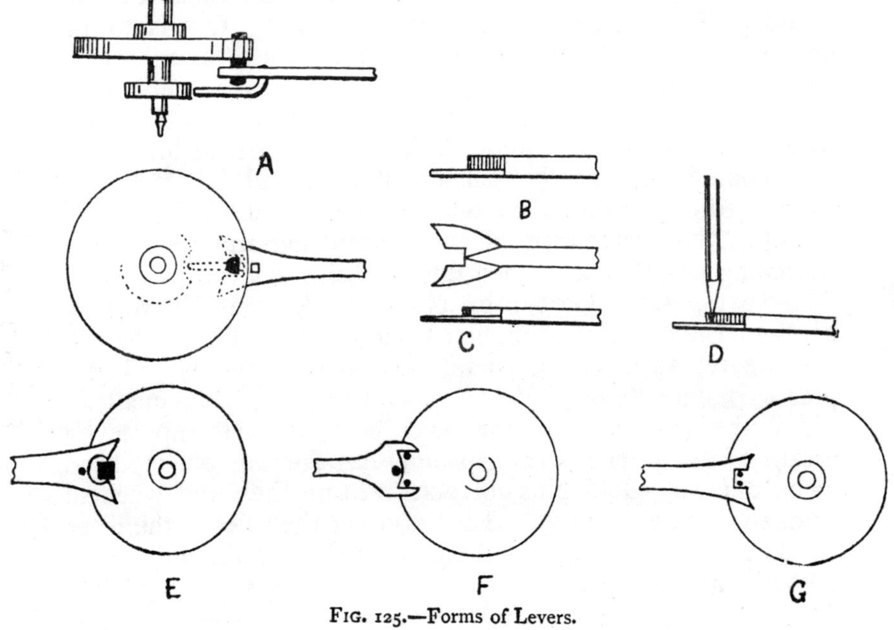

FIG. 125.—Forms of Levers.

impulse, a hollow is cut in the roller edge, opposite the ruby pin.

This escapement has an ordinary lever and roller action, but there are many variations. A "double roller" is shown in Fig. 125 at A. In this there is no hollow in the roller edge, and no guard pin in the lever; but, instead, the lever has a "dart" affixed to it, which acts in conjunction with a second small roller underneath the large one. This is a very safe form, and much to be preferred to the ordinary "single roller." As shown in Fig. 125, the "dart" is a bent wire riveted into the lever. In English watches the dart is generally of gold and fixed to the lever by a screw and steady pin. In the best Swiss watches also the dart is fixed in this way, but is of steel and very light. It may be mentioned that the little screw is the smallest made and used in watchwork, and probably the smallest screw used for any purpose.

At B (Fig. 125) a common Swiss form of lever is shown. There is no guard pin, but, instead, the upper part of the lever is filed up to form a dart. When these escapements require the dart advancing to make the action safe, they cause a lot of trouble. There are three methods of doing it. One is to file back the dart point and soft solder a new brass "nose" on it, filing up to shape, as at C; or the dart may be slit with a screw-slitting file, and the end forced forward by a chisel-shaped punch being hammered into the slit, as at D; or the lever may be softened between the pallet-staff hole and the fork, and hammered on its sides to stretch and lengthen it. Before doing this take its measure with the millimetre gauge and see how the stretching progresses. When done polish the lever up again.

Sometimes a square ruby pin is used, as at E, working between the edges of a circular lever notch; or the ruby pin may be replaced by two metal pins, as at F and G. These pins are generally of gold. F is a "two-pin" escapement, in which the guard pin in the lever helps to give impulse by means of the roller notch. G simply has the ruby pin replaced by gold pins. Gold pins get worn into notches, and when so worn should be replaced. They also cut the sides of the lever notch, and necessitate smoothing and polishing the notch with oilstone dust and red-stuff.

Faulty Escapements.—Lever escapements may have many faults, some of construction, some caused by careless bending of banking pins, etc., and some by wear. A most serious fault is "mis-locking." When this fault is present the points of the scape-wheel teeth, instead of falling just on the locking faces of the pallets, as at A (Fig. 126), miss the corner and fall upon the impulse faces, as at B. This is caused by the points of the scape-wheel teeth wearing short, or by the scape or pallet pivot holes wearing wide, allowing the scape wheel and pallets to get too far away from each other. A very little is sufficient to cause a locking escapement to mis-lock. A watch that mis-locks is sure to stop. To test for this, lead the balance rim slowly round with the finger tip, while a little pressure is applied in a forward direction to the scape-wheel teeth with a flat-cut peg point. Watch it give impulse until a tooth "drops." At this moment

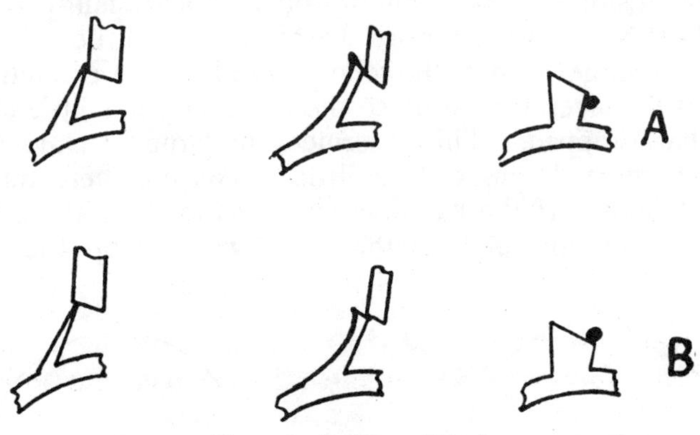

FIG. 126.—Locking of Teeth.

let the balance go free. If it has locked, the scape wheel will remain stationary and draw up the lever smartly to its banking pin. If it has mis-locked the lever will go back and the scape wheel will go forward, giving another impulse in the reverse direction. An escapement should be tried thus before taking the watch to pieces, so that, if faulty, it may be corrected.

American watches, and some English and Swiss, have sight holes in the plate, through which the pallet depth can be easily observed without having to test as above.

To correct a mis-locking escapement, if due to wide pivot holes, bush them; if due to worn teeth, bush the top pallet hole, drawing it a little nearer to the scape wheel, as shown in Fig. 88, p. 83.

Frequently a lever watch is found stopped with the lever on the wrong side of the roller. This is caused by the guard pin being too far from the roller edge and allowing the lever to pass when shaken. To remedy, the guard pin may be bent a little forward, until it will not pass or stick against the edge. A guard pin that will not quite pass, but sticks, is sure to stop the watch, and a watch that is found to stop with the slightest movement—taking out of the pocket to see the time, etc.—very likely has this fault. A slight roughness on the roller edge will also cause the guard pin to catch and stick.

Swiss watches with solid "darts," as in Fig. 125, at B, are more troublesome, and the methods of advancing them are described on p. 104.

The position of the banking pins is often faulty. They should be wide enough apart to allow of a little "banking shake" on either side of the roller, and also wide enough apart to allow the lever to run up to the banking pin a little after a tooth has dropped. This is termed the "run." Only a little "run" is wanted, and only a little shake, but there must be some of both. If the banking pins are too close there is apt to be no run on some teeth, which would stop the watch

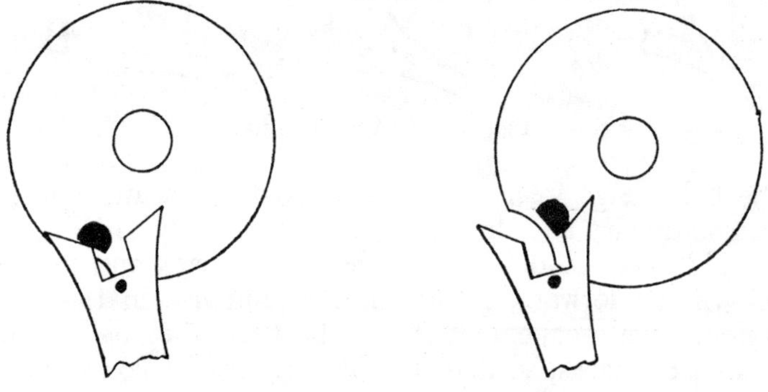

Figs. 127 and 128.—Faulty Lever and Roller Actions.

instantly, or not sufficient banking shake, which would cause the guard pin to touch the roller edge; or it might cause the face of the ruby pin to butt against the corner of the lever notch in entering, as in Fig. 127. If too far apart the "run" will be excessive, causing resistance to unlocking and a bad action; or the ruby pin may butt against the other corner of the notch, as in Fig. 128. A watch that has a bad action when hanging up, with the balance leaning towards the lever, will generally be found faulty in the lever and roller action. If it has a round ruby pin, change it for one with a flat—D shaped—as shown in the figures. To test for these little faults, insert a wedge of pith or cork under the balance rim to keep it steady. Lead it round with the finger tip very slowly, and with tweezers try the shake or freedom of the lever at each point. Any tight spot, caused by fouling, will be found at once.

American watches, and many machine-made English and Swiss also, have adjustable banking pins, which are a great convenience. These are usually screwed into the watch plate, and the pin is eccentric. A slight turn of the screw head serves to move the pin as desired.

The ruby pin should have just a little side play in the lever notch, and not fit it tight. A flatted pin may be made from a round one by taking a piece of brass and filing a groove across its face, deep enough for the pin to lie in with one quarter of its thickness standing above the surface of the brass. Cement the pin in the groove with shellac. Screw the brass in the bench vice, and flat the pin down with a $\frac{3}{0}$ emery buff.

It may be mentioned here that "ruby" pins, like "jewel" holes and pallet stones, may be *glass* in very common watches, are *garnets* in most, and *ruby or sapphire* in a few of the very best. A ruby could not be thus cut with emery buffs, but glass or garnet can.

To cement a ruby pin in a roller, a roller-holder, like Fig. 129, will be wanted. This is to warm the roller upon to run

FIG. 129.—Roller Holder.

the shellac. Holding the roller in or over a flame would probably blue it, and in any case the roller would not remain hot long enough to enable the pin to be set true. The holder is made of stout brass wire 4 in. long, filed smooth, and tapered at one end for about 1 in. At 1 in. from this end a circular piece of brass is driven on tight, to act as a heat retainer. To use it the roller is pushed on the tapered end, the brass block A is held for a moment in the spirit lamp flame and then removed. Enough heat is retained to keep the roller hot. The remains of the old broken pin are then pushed out with a peg. Still holding the roller on the tool, select a new pin to fit, put it in position, and pick up a small piece of flake shellac. Warm the tool again, and apply the corner of the flake of shellac to the back end of the ruby pin until a little melts and runs on it. Work the pin backwards and forwards a little in the hole, and then, while the shellac is half set, set the pin upright with tweezers, and see that the flat is outwards. When cold, cut off the surplus shellac from the pin sides and the roller face and edge with a sharp graver.

In handling a roller—to take it off a balance staff or put it on—use only brass-nosed pliers. Steel pliers will rough its edge and cause the guard pin to catch. A tight roller is best removed with a "roller remover," a useful little tool with a lever action, to draw them off without injury.

Many common watches, especially some English full-plate fusee levers, have pallets not suited to the scape wheel. If the pallets are too small for the wheel they will also be too close together, and the "inside" drop of the teeth will not be sufficient. The scape teeth must have *some* drop on to the pallet locking faces, or else the point of the pallet will scrape down the back of the tooth. With pallets too small for the wheel, there will not be enough drop upon the inside locking face, and the point of the short or "entering" pallet will scrape the backs of the teeth. If the pallets are too large there will not be enough drop on the outside locking face, and the point of the other pallet will foul the teeth backs.

This fault causes a bad action, and can be detected by putting the lever and pallets and scape wheel in a depth tool and adjusting to a correct depth. The action can then be leisurely observed. To remedy, a little must be taken off the faulty pallet corner with a $\frac{3}{0}$ emery buff, or with flour emery and oil on an iron lap in a lathe, or with a "ruby file." It may here be mentioned that English watches have the pallets simply driven tight on to the pallet staff, which is tapered for that purpose, and can be removed by driving the staff out with a punch. The lever is simply pinned on to the pallets with two brass pins, which can be pushed out from underneath. Swiss and American watches have the lever and pallets screwed on to the pallet staff, and can be removed by holding the pallet staff in brass-lined pliers and unscrewing them with the fingers.

English watches generally have "covered" pallets, like Fig. 120, the pallet stones being sunk flush in slits made to receive them. They are cemented in with shellac, and can be re-cemented or moved by warming a pair of pliers and holding them while the stone is adjusted. Swiss and American pallets are nearly always "visible," like Fig. 121. To re-cement or move these stones is easier. Warm a brass plate and clamp the pallets down to it; then operate on the stones. These visible pallets are more fragile and liable to accident than covered ones, and often become loose or chipped. To adjust a pallet stone is a delicate operation. If one has become loose, re-cement it and push it home in its groove. Then try the escapement in the watch frame to see if it locks properly. If the stone wants advancing a little, first take its measure with the millimetre gauge, and the amount it is moved can be determined with accuracy.

If it is desired to try a lever and roller depth in the depth tool there will be a difficulty in getting the balance in. The roller should be taken off and placed on a true turning arbor, and this put in the tool.

To test the "draw" on the pallets, with the tweezers or a peg point, move the lever a little away from one banking pin, but not enough to unlock the wheel. If there is "draw" the tooth point will draw the pallet up again, and the lever will return sharply to the pin. If there is insufficient draw the lever, instead of returning sharply, will move sluggishly or stop still where moved. This should be tried on each pallet, that is, against each banking pin. Want of "draw" causes the guard pin to drag on the roller edge between the beats and makes the action poor. Pallets with no draw can only be properly corrected by a pallet maker. Pallet making is a trade to itself, and is one of the many subdivisions of labour in the making of a watch. Some repairers give "draw" to a pallet by shifting the stone a little aslant, but this upsets other adjustments, and is not to be recommended.

Bent scape-wheel teeth can generally be straightened with tweezers by the eye as a gauge. To make sure, put pallets and wheel in the depth tool and examine the drop before putting the watch together.

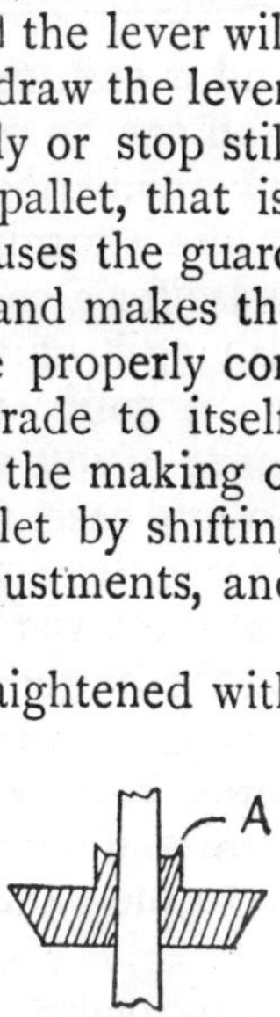

FIG. 130.—Shape of a Wheel Seating.

Unlike the other wheels of the watch train, the scape wheel is mounted on a brass collet or seating, instead of being riveted on the pinion. The collet is driven tight on the pinion, and then turned to receive the wheel, as in Fig. 130. A thin riveting edge, A, is left standing above the wheel surface. This is not turned over with a punch, but is burnished over as the wheel runs in the turns or lathe by a fine burnisher with a smooth, rounded point, lubricated with oil. If a scape wheel is found to be loose on its collet, re-burnish it a little.

To turn in a new pallet staff will require no special directions; a piece of steel wire being centred, hardened, tempered blue, turned taper to fit the pallets, polished and pivoted as described on p. 87.

A balance staff is not such a simple job. They are of two kinds (Fig. 131). A shows a rough "brass-collet" staff as obtained from the tool shop, and B a rough "solid" staff. Most repairers use these rough bought staffs, though many make their own. The brass-collet staff A is a piece of hardened and tempered steel ready centred for turning, with a brass collet driven on to it. The solid staff is made of one piece of steel and is soft, requiring hardening and tempering. A brass-collet staff can be made by taking a short length of steel wire, hardening, tempering, and centring it, turning it tapered, and polishing it like a turning arbor, to fit the roller; then drilling a piece of brass for a collet and turning it upon a separate arbor, finally driving it on the polished staff to the correct height.

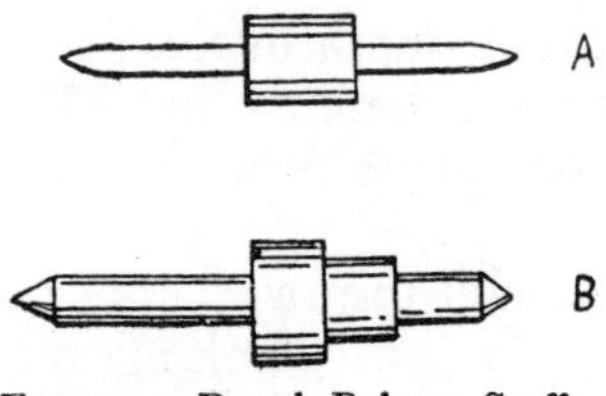

FIG. 131.—Rough Balance Staffs.

A solid staff is made by turning roughly to shape from a piece of soft steel rod, hardening and tempering, then finishing. In any case, the first thing to do is to rough down the collet to size, leaving it just a little too large everywhere. Then to cut the lower part of the staff to length, so that when stood up on the bottom endstone (hole removed) the balance seating can be sighted at the correct height. Then turn and polish the lower part of the staff, tapered to fit the roller, seeing that it also comes at the right height to work with the lever. Turn and polish the bottom pivot to fit the jewel hole and round its end up. The staff can then be reversed and the top end cut off to length. The total height can be measured by filing a piece of brass to stand between the endstones, with jewel holes removed; or can be measured over the outsides of the holes, with endstones removed, by douzième or millimetre gauge.

The balance seating can next be turned flat and true to

receive the balance *tightly;* the riveting face must be turned hollow to leave a nice thin edge for riveting over the balance, like A (Fig. 130), the hairspring collet fitted a nice sliding fit, and the surplus metal of the staff collet cut off, leaving the upper part of the staff from which to turn the top pivot. This can then be turned, polished, and rounded up. Finally, rivet on the balance. Pass the staff through a hole in the steel stake, so that the collet rests on the stake. Use a crescent-shaped punch (Fig. 132), and tap round and round gently until the riveting edge is well down. Then use a flat-faced punch with central hole to finish, and turn off the riveted face smooth.

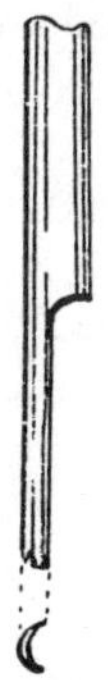

FIG. 132.—Crescent or Half-round Riveting Punch.

True the balance in the way described on p. 84 and "poise" it. If a watch is to keep time, the balance must be in perfect poise, that is, it must have no heavy part and show no tendency to settle at any one point when placed in a pair of calipers. A balance poising tool consists of a pair of parallel straightedges that can be adjusted to suit various balance staffs. The pivots are rested upon these, and any heavy part will at once settle at the bottom. To poise a plain gold or steel balance, file a little off the inside edge of the rim at the heavy part. A compensation balance can be poised by its quarter screws, or if out too much for that, small washers (sold for the purpose) may be put under the screw heads, or screws can be slightly reduced.

A balance staff has cone-shaped pivots for strength. These can be turned by a thin graver with a rounded point, rubbed round on the oilstone. The polisher must be curved in section to suit them, as in Fig. 133, C. They should be smoothed with oilstone dust until the ends just go in the jewel holes, and polished with red-stuff until the ends come well through, as at A (Fig. 133). Although called "cone" pivots, the part that goes in the jewel hole must be straight; only the shoulder is curved, for strength. A pivot of this shape will naturally draw the oil away from its point and let it spread up the staff. To prevent this, cone pivots are "back cut." An enlarged "back cut" pivot is shown at B.

In the turns a balance staff is manipulated exactly as

described in turning pinions. In a watch lathe the staff is roughed out in split chucks, then cemented in the shellac chuck (Fig. 100, p. 89), and the entire lower part finished; then taken out, measured, and cut off to total length, and reversed

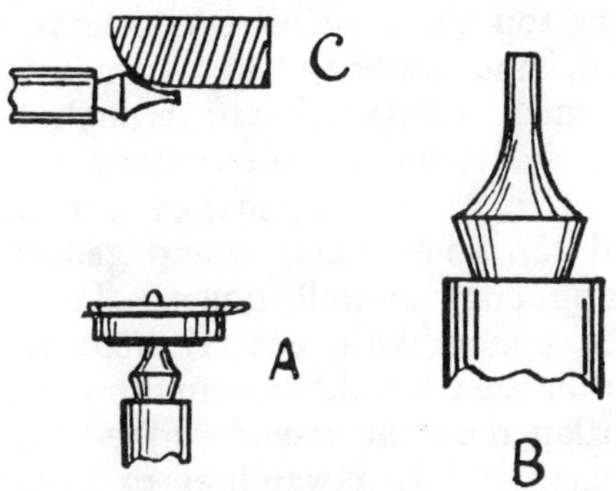

FIG. 133.—Polishing Cone Pivots.

in the chuck, the balance seating, etc., and top pivot being done last. The whole is done in two chuckings.

Fig. 134 shows an undersprung and an oversprung staff. A = seating for hairspring collet. B = balance seating.

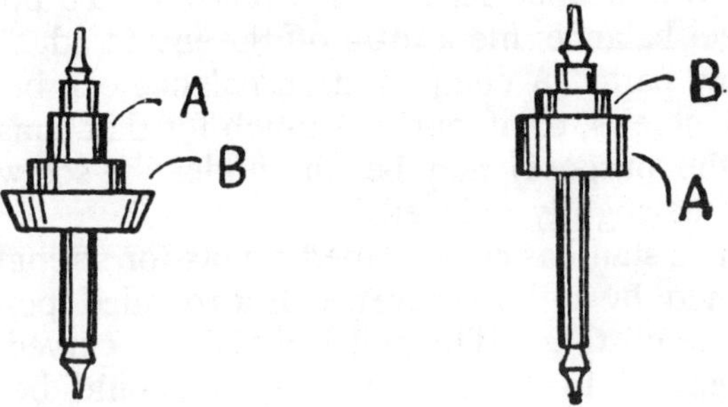

FIG. 134.—Undersprung and Oversprung Balance Staffs.

Bent balance-staff pivots are straightened with brass-nosed pliers. Sometimes, when a watch has had a blow, the end of one pivot is riveted like a miniature mushroom against the face of the endstone, as at A, Fig. 135, and cannot be withdrawn from the jewel hole. In such a case something is bound to smash. The pivot must be pulled out by main force and may part, or the jewel hole may break. A broken pivot means a new staff. A pivot bent like a dog leg, as in Fig. 135, B, is no good. It cannot be got right, and a new staff is needed.

Sometimes a blow only slightly bruises the end of a pivot, dumping it up shorter. A little polishing will put such a pivot right, and it should be tried to see that it comes through its jewel hole properly. A cone pivot can be made to come further through its hole by polishing the curved shoulder back a trifle. A staff, the pivots of which have been treated thus, is shorter than it was before, and will sometimes let the balance down so low that it touches the plate or scape cock. A little can generally be done by getting the bottom jewelling up. Take out the jewel hole, and, laying it on the finger tip, file the surplus brass over its top to let the endstone lie as close as possible to the hole itself. There is usually a little

FIG. 135.—Damaged Pivots.

that can be removed. Also file away the surplus brass on the endstone face in the same way. If sufficient room cannot be got in this way, there is no remedy but a new balance staff, unless the staff is made by driving a brass collet on to a tapered and polished staff, as described on p. 110. Then the staff may be driven a little further through the collet and the balance thus raised. But very few balance staffs are now made in this way. They are only found in London hand-made watches.

The lever escapement is found arranged in many ways in different watches. In Swiss and American $\frac{3}{4}$-plate watches the scape wheel is frequently under the top plate with the train wheels, the lever being held by a small cock sunk under the balance rim. At other times the scape wheel and pallets are under one "scape cock," as in English $\frac{3}{4}$-plate watches. Full plates, both English, Swiss, and American, have the balance carried by a "potance" under the top plate, the scape wheel and pallets being in the frame with the train wheels. A few English full-plate watches have the balance staff jewelling in the pillar plate, and the balance lies in a sink or hollow turned in the top plate. This plan enables the hands to be set from the back, and does away with the potance. For some reason not very apparent, watches made thus are seldom very satisfactory. They have a bad action, are hard to see and examine properly, and generally prove unsatisfactory to their wearers.

Further information on the construction of the lever escapement will be found in the chapter on "conversions."

Cylinder Escapement.—This form of escapement is going out of use rapidly. Probably in the near future it will only be found in some Swiss ladies' watches, for which it answers fairly well.

Fig. 136 shows the action and the different parts. A = cylinder, B = scape wheel. It consists of a scape wheel with fifteen peculiar teeth and a hollow steel cylinder, a part of which is cut away. Each tooth of the scape wheel is mounted upon an upright stem, standing up from the scape-wheel rim. On each tooth is an inclined plane or nearly plane curve. The central part of the cylinder, where the scape-wheel teeth engage it, is cut away, leaving just a little more than half the circle. The wheel and cylinder are planted so that the centres of the inclines on the scape-wheel teeth pass through the centre of the circle of the cylinder. Suppose a tooth rests inside the cylinder as in the figure at A. The point of the tooth presses against the inside surface of the cylinder shell and the wheel cannot advance. As soon as the balance and cylinder turns in the direction of the arrow the tooth point will be released, as at C, and the scape wheel will advance, the inclined face of the tooth giving an impulse to the cylinder edge or "lip." When the tooth has quite passed, the next tooth will fall against the outside surface of the cylinder and lock the wheel once more. The cylinder returning, the point of this tooth will be released and it will give impulse as at D, until its point drops against the inside surface of the cylinder shell, and so on.

This is really a very simple escapement; but because it is somewhat hidden in the watch its action is much misunderstood, and is a great mystery to many beginners. The cylinder must be of such a size that it passes between two teeth of the scape wheel with a little to spare, as shown at E. Also it must be large enough for a tooth to be completely inside it and have a little shake, as at A. The amount of shake gives the "drop" of the teeth. Thus a cylinder too large for the wheel will have no outside "drop," while one too small will

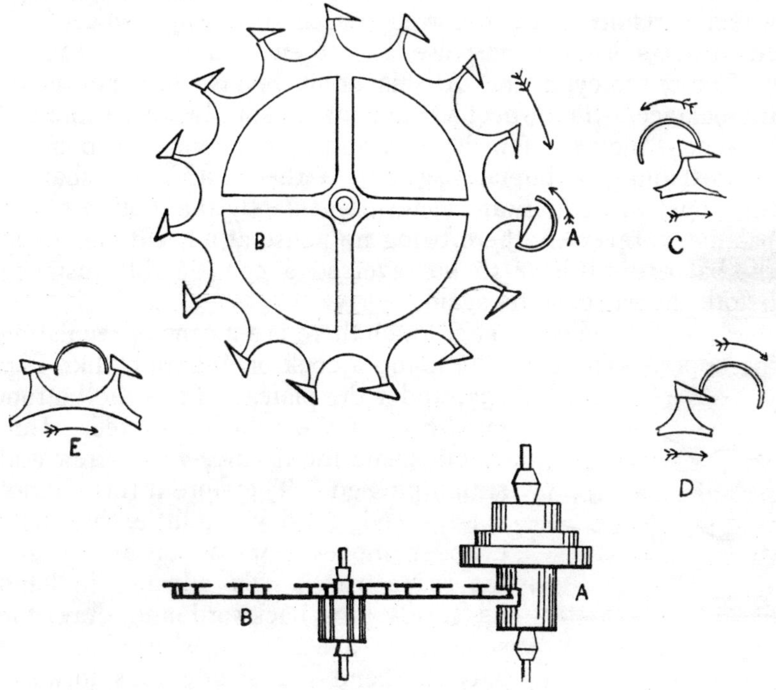

FIG. 136.—Cylinder Escapement.

give no "drop" to the teeth inside. In a correct escapement the inside and outside drop should be equal. A watch that suffers from too little drop, either inside or outside the cylinder, may be made correct enough to go by filing the least trifle off the points of the teeth, shortening them a little. They must all be done equally, and not one missed.

The depth of the cylinder and scape wheel is not a very particular one within small limits. But if too shallow it will "run through." This is caused by the points of the teeth missing the surface of the cylinder and getting past its edges, thus giving impulse as soon as they drop. A watch with this fault will tick rapidly and unevenly. Too deep a depth is shown by a want of shake of the teeth inside the cylinder and by the balance having to turn a long way round to unlock the teeth points.

The depth can be observed by leading the balance rim round with the finger tip and watching the motion of the scape wheel carefully. In a correct depth the scape wheel will advance as it gives impulse, and then "drop" on to the surface of the cylinder. At this point reverse the motion of the balance. The wheel should stand still, locked, while the balance is moved a few degrees, and then commence to move forward and give impulse again. If the depth is too shallow the wheel will go forward again immediately the motion of the balance is reversed, there being no pause at all. If too deep the balance will have to be reversed a considerable distance before the wheel starts again.

In nearly every cylinder watch there is a means of regulating the depth. There will be found a cock or "chariot," like Fig. 137, under the plate. In this the bottom jewel hole of the cylinder is fixed. Into its other end the balance-cock screw and steady pins go. Therefore if this chariot is advanced, or backed a little, the whole cylinder moves towards or away from the scape wheel. To advance it, bend its steady pins backward and draw the screw hole a little oval with a file. To back it, bend the steady pins forward. It may be necessary, to enable the chariot to be moved, to trim a little off some of its edges or off the plate by filing.

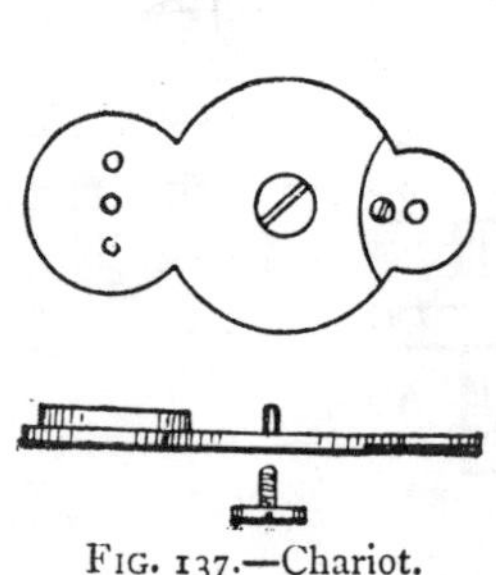

FIG. 137.—Chariot.

Another important point in this escapement is the relative height of the cylinder and scape wheel. The scape wheel should pass through the "passage" cut for it in the cylinder without touching either the top or the bottom of it, as shown in Fig. 136. When it touches either the top or the bottom badly the watch will stop; and while going the scape wheel will be seen to "dance" at each beat. When held to the ear, first dial up, then dial down, a knocking or scraping will be heard. Too much endshake of the cylinder or scape wheel will cause fouling here. The endshake of both wheel and cylinder should be very little, and be equal. They will then not alter their relative positions when the watch is turned over. A cylinder that is too high can be got down by putting a small slip of paper under the chariot. One too low is not quite so easily got up. If the bottom endstone is set in a brass slip, the cylinder can be raised a little by breaking the endstone out, chamfering its setting out, and fitting a loose endstone as thick as it will take. This saves the room that was wasted between the endstone and the jewel hole. Or the face of the chariot may be filed so that it will lie closer to the under side of the plate.

In the rim of the balance there will be found a little pin. This is to prevent the balance making more than half a revolution in either direction from its point of rest. It should come in contact with a fixed pin in the back of the balance cock. If the pin in the rim is too short or is lost, and the balance makes more than half a turn, it will be caught by the scape wheel teeth and held firmly from returning. Or if the pin is in the wrong place, allowing the balance more than half a turn in one direction, the same thing will occur. In some watches the pin in the rim banks against the fourth pinion instead of a pin the back of the cock. The pin in the rim should be in such a position that, when the watch is in beat and the balance at rest, it is exactly half a turn from the pin it banks against.

The cylinder itself is a thin steel tube. Its ends are closed with turned steel plugs driven in tight. Upon these plugs the pivots are turned. Upon the outside of the tube a brass collet is driven, and is turned to form a seating for the balance and the hairspring. Cylinders in the rough—that is, with the tube finished and polished, the plugs fitted, and the collet on—can be purchased for a few pence. They require cutting down to the correct length, pivoting, and the balance and hairspring fitting.

To turn in a new cylinder is not such a difficult job as many beginners think. In a watch lathe the cylinder can be held in the cone cement chuck, like a balance staff, and the lower portion turned and pivoted, the cylinder body being filled with shellac to stiffen it. It can then be taken out, measured for correct height, and reversed in the chuck for finishing the balance seating and top pivot. In the turns the cylinder must also be filled with shellac, as in a lathe. To do this, place a flake of shellac inside and warm it gently over the lamp flame—not *in* the flame, or it will be softened. A brass

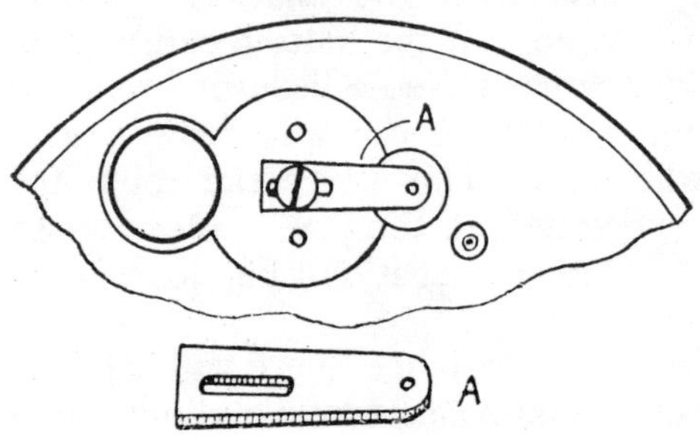

FIG. 138.—Temporary Chariot.

ferrule, with a plain circular hole in its centre, is cemented on the cylinder with shellac. Whether this ferrule runs exactly true or not does not matter. The pivots are turned and polished, etc., as before described on p. 86.

The principal difficulty is to get all the heights right. There are cylinder height tools sold, but they are not really necessary. The simplest process is to remove the under

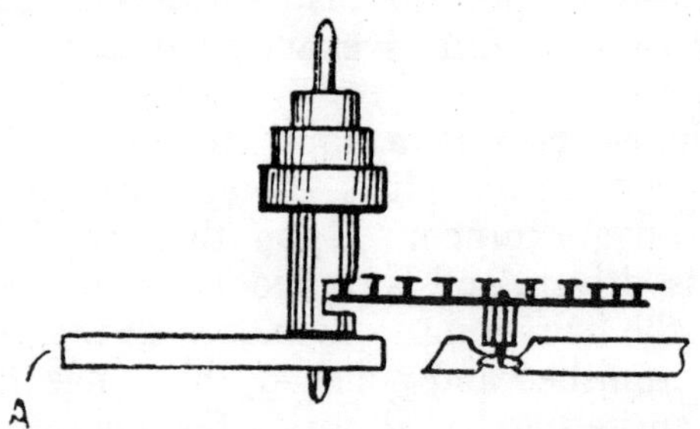

FIG. 139.—Adjusting Height of a New Cylinder.

chariot from the plate, and in its place screw the brass piece A, Fig. 138. This piece can be adjusted so that the pivot hole comes about the right place. The hole should be large enough to take the cylinder plug and allow the body to stand upon it, as in Fig. 139. It can then be seen if the passage in the cylinder comes at the right height for the scape wheel. The bottom of the cylinder body must be turned away until it is right, and a *little more.* Then turn the bottom pivot, leaving it full long, shorten it down until when in its jewel hole it is right height. Next sight the level for the balance so as to just free the scape cock. Turn this and the shoulder for the hair-spring collet. Finally, measure the total height of cylinder with douzième or millimetre gauge, over the outsides of the jewel holes with endstones removed, and turn the top pivot. In riveting the balance on, care must be taken to see that the banking pin comes in the right place, as explained on p. 117.

A bent cylinder can rarely be straightened without breaking, but it may be tried with brass-lined pliers, springing it carefully. Bent pivots can generally be managed. A broken pivot is replaced by knocking out the cylinder plug and fitting a new one, turning a fresh pivot upon it. Cylinder plugs can be purchased ready made, but are not necessary if the workman has a lathe. They are so easily made.

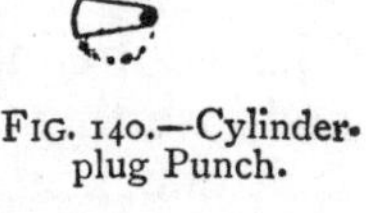

FIG. 140.—Cylinder-plug Punch.

To knock out a cylinder plug is sometimes very difficult. The cylinder is first rested upon the brass collet, so as to give a good seat, and the punch shown in Fig. 140 used to drive it out. If the plug is tight the cylinder may be driven through the collet, instead of the plug moving. In such a case, rest the cylinder end over a hole in the stake which the plug will go through and the cylinder will not, and try again. Or a stake with coned holes may be used, as in Fig. 141, which allows the plug to be just started. When a plug has been knocked out just a little and is still tight, it need not be further touched, but the new pivot may be turned upon it. A plug that cannot be moved by any means may, in a watch lathe, be easily drilled, and a pivot put in as if it were a pinion. A "plug" is shown in Fig. 140 at A.

A scape wheel bent out of flat may be made true by bending with tweezers, springing it carefully in the desired direction. Broken teeth cannot be properly replaced, but require a new scape wheel. New wheels cost a few pence each, and can be purchased the exact size required, ready to put on the scape pinion. The scape wheels are not riveted on the pinions, but only driven on tight a good fit, and may be easily removed by punching the pinion through.

A new scape wheel will require opening out in the centre to fit on the pinion. As bought they are too hard to broach.

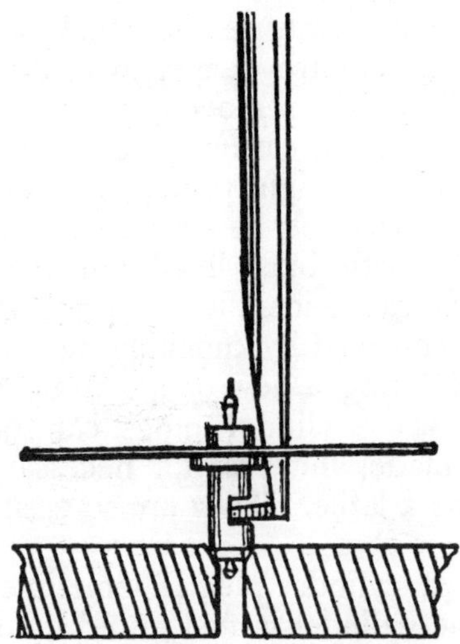

FIG. 141.—Removing a Cylinder Plug.

The centre may be softened by heating a tapered brass wire and inserting into the hole. Or the pinion may be held in a split chuck in the lathe and turned down to fit the wheel.

A broken scape pivot can be put in with a watch lathe by first knocking off the wheel, then putting the pinion in a split chuck and drilling. A pinion with both pivots off may be drilled right through and a piece of steel driven in, a pivot being turned on either side. This job in the turns is a nearly impossible one.

The tops of the scape-wheel teeth sometimes run foul of the scape cock. To free them, the cock must be cemented flat on a piece of sheet brass with shellac and put in the mandrel; or it may be cemented on to a flat-faced cementing chuck in a watch lathe, and the passing hollow turned out a little more.

Old English cylinder watches have brass plugs to the cylinder, in the centre of which steel inner plugs are driven to form the pivots upon. These watches also often have brass scape wheels. This causes the cylinder sides to wear badly. A new cylinder to one of these watches is not a very formidable job. A piece of steel rod is drilled, and smoothed inside by a brass rod and oilstone dust and red-stuff. The plugs are fitted and the collet driven on. The outside is then polished and the body cut open.

Verge Escapement.—This escapement requires very little explanation. It is now nearly obsolete, and, it is hoped, will be quite so in a few more years.

The scape wheel A, Fig. 142, has thirteen or fifteen saw-

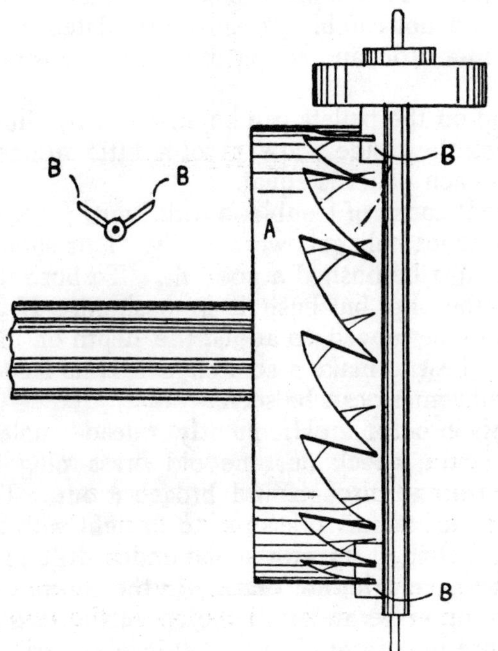

FIG. 142.—Verge Escapement.

shaped teeth. It runs vertically in the watch, the back pivot running in a "follower," and the front or "inside" pivot in a brass slip in the potance face (see Fig. 71, p. 66). The teeth points act alternately on the pallets BB, which are flat steel "flags" projecting from the slender verge body. In this escapement the balance is never free from the scape wheel; when the teeth are not driving the pallets, the pallets are causing the teeth to "recoil." Therefore any variation in the motive power directly affects the time of the watch, and a "fusee" is a necessity. Also a stronger mainspring will cause the watch to gain, or a weaker one will make it lose, which is not the case with either a lever or a cylinder watch.

It may have, and generally has, many faults. The depth may be too shallow, in which case the "drop" on both pallets will be excessive, and the balance will perform very short arcs, causing the watch to gain. If too deep the teeth points will catch on the verge body, or there will be no "drop," and the watch will stop. This depth should be set as deep as possible for it to go and not catch. It can be regulated by driving the "follower" in a little, and so getting the scape wheel nearer to the verge.

The drop on the pallets can be adjusted by sliding the slip which carries the inside scape pivot a little to one side, until the drop on each pallet is equal.

A frequent cause of trouble is wide worn pivot holes. The inside scape pivot hole gets worn badly. This should never be passed, but must be bushed a good fit. To bush this hole, do not remove the slip, but bush it in position. If at any time the slip has to be moved to adjust the depth or for any other purpose, it is best to make a scratch across its face so that the amount of movement can be seen.

Verge pivot holes are frequently "dead" holes in brass. To fit new ones, knock out the old brass plug in cock or potance, or drill it through and broach it out. Then file up a good flat-ended round brass pin to fit tight with the flat face flush inside. Drill it up with a flat-ended drill to a sufficient depth, a little eccentric, not exactly in the centre of the brass. Then put in the verge and turn the eccentric brass plug round until the verge is upright. Cut it off level and rivet in.

To properly re-turn and polish an inside scape pivot that is badly worn, the scape wheel must be knocked off. When riveted on again it must be topped true by revolving rapidly in the turns and holding a pivot file to the teeth points until it just marks all of them. They must then be filed up to good points by operating on the *curved backs* only with a fine crossing file, the wheel being held in a pin vice by its pinion.

Verge pallets that are cut into holes by the teeth may be

smoothed down and polished with oilstone dust on a flat steel polisher, followed by red stuff.

A little fault often found in these escapements is that the top of the scape wheel touches the under side of the brass collet of the hairspring.

In a watch lathe a new verge is put in by cementing in a cone chuck to turn the collet and pivots. In the turns, the slender verge body is stiffened by cutting a short piece of watch peg as in Fig. 143. The verge is placed in the hollow and the loose piece fitted in, pushed tightly into a solid brass ferrule and shellaced firm. It can then be turned and pivoted. The pallets must be shortened off by filing to make the escapement drop off properly, and the backs bevelled off to a knife edge.

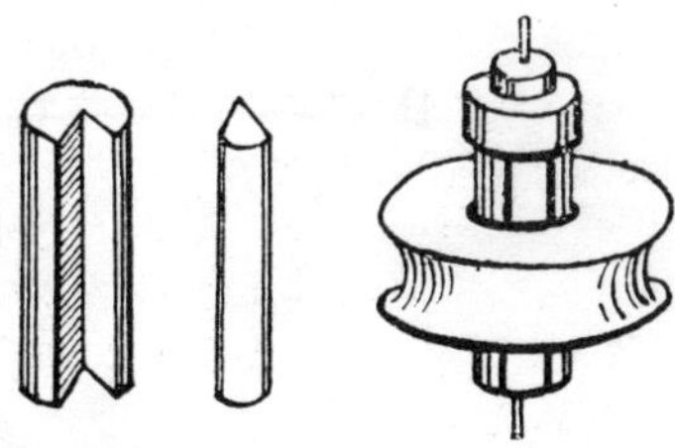

FIG. 143.—Turning Verge Pivots.

A verge balance has an upright banking pin in its rim, which banks against two fixed studs, or two shoulders in the balance cock. When in beat, and the balance at rest, this pin should allow the balance to turn an equal distance in either direction.

Duplex Escapement.—This is not now made, with the exception of "Waterbury" watches, which still have it. Generally a good old English watch with this escapement pays to convert to a lever.

Fig. 144 shows the action. The scape wheel A has two sets of fifteen teeth. One set are long and pointed (D), the other are short and upright (C). The teeth D are for locking the wheel, and rest upon a ruby roller, E, on the lower part of the balance staff. This roller has a vertical slit, which just allows the teeth points to pass as the balance comes round. The upright teeth C are to give impulse, and, just as the ruby roller allows a long tooth to pass, one of the teeth C gives an impulse to the point of the long pallet B on the balance staff.

The pallet B is fitted on to the staff like the roller in a lever escapement, and can be turned round as desired to adjust the drop. The ruby roller E is a tube of ruby slit down one side, and is slipped a sliding fit on to the lower part of the balance staff and cemented with shellac. A small brass cap, F, is pushed on after it and cemented also.

With wear the points of the teeth D shorten, and this, combined with wide pivot holes and worn pivots, causes the teeth to slip by the ruby roller. The remedy for this is to repolish

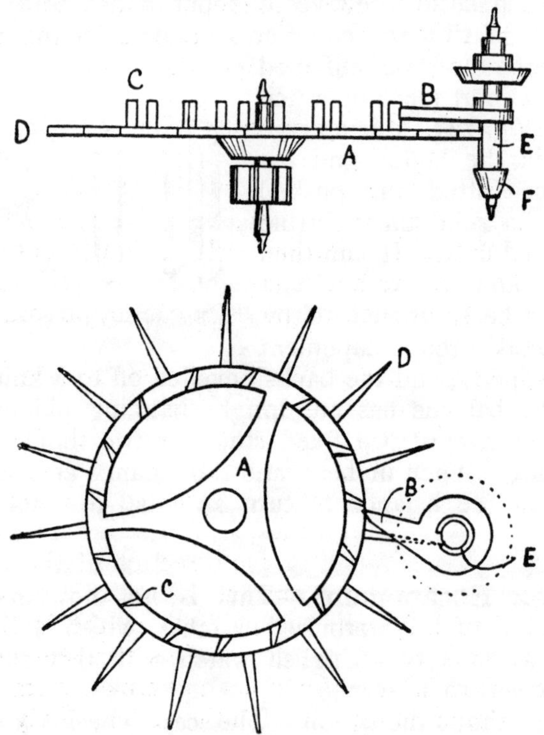

Fig. 144.—Duplex Escapement.

the pivots and fit new jewel holes, drawing the jewelling a little if necessary. The drop of the short teeth upon the pallet B should be quite perceptible on each tooth, and must be carefully noted by leading the balance round with the finger tip until a long tooth escapes and watching the amount the wheel moves before a short tooth falls upon the pallet. To give more or less drop, the pallet B can be turned round.

In old watches the long teeth points wear off, and the short teeth cut until they nearly miss the impulse pallet. Such a watch wants a new scape wheel. A broken balance staff and a cracked ruby roller are also the frequent results of a fall. Any of these repairs are troublesome and expensive, and, if required, it will generally be found best to at once convert the watch into a lever.

In Fig. 144 the point of a long tooth is shown just about to escape from the slit in the ruby roller, and a short tooth is about to "drop" on the pallet B and give impulse. No impulse is given on the return vibration.

Chronometer Escapement.—This was invented for marine timekeepers, to keep Greenwich time on board ship for the purpose of ascertaining the longitude. It is a very delicate escapement, but is capable of keeping correct time for a long period without much variation. It is sometimes found in pocket watches, and, in the hands of careful wearers, is fairly successful. Still, for such purposes, it cannot be said to equal a fine adjusted lever.

Fig. 145 shows its arrangement, as made in England. The scape wheel A has fifteen teeth of the shape shown. The wheel is locked by the teeth falling upon a locking stone (ruby), C, set upright in a spring detent, B, which is screwed to the watch plate. The balance staff has two rollers upon it. The small one has a ruby pallet, E, projecting from it, which, as the balance comes round in the direction of the arrow, will lift the point of the detent B and let the tooth H escape. The tooth I will then fall upon the pallet G in the large roller F and give the balance an impulse. This escapement, like the duplex, only gives an impulse at each alternate vibration of the balance. On the return vibration the pallet E only raises the thin gold passing spring D, and does not move the detent itself.

In this escapement the balance is more free than in any other, and it requires no oil upon the scape-wheel teeth or on the detent point.

Both rollers can be moved upon the balance staff for adjustment. As in the duplex, as soon as the detent is raised and a tooth, H, escapes, a tooth, I, should have a little "drop" on to the pallet G. When the wheel is locked by the detent, the large roller F should be just free of the points of the two scape-wheel teeth between which it revolves.

The detent banks up against a screw stud, and its exact position can be regulated. To be correct, the gold spring should point exactly to the balance centre. The slightest trace of oil upon the point of the banking screw, or on the detent point, where the gold spring rests upon it, will make these parts sticky and cause great irregularities of rate.

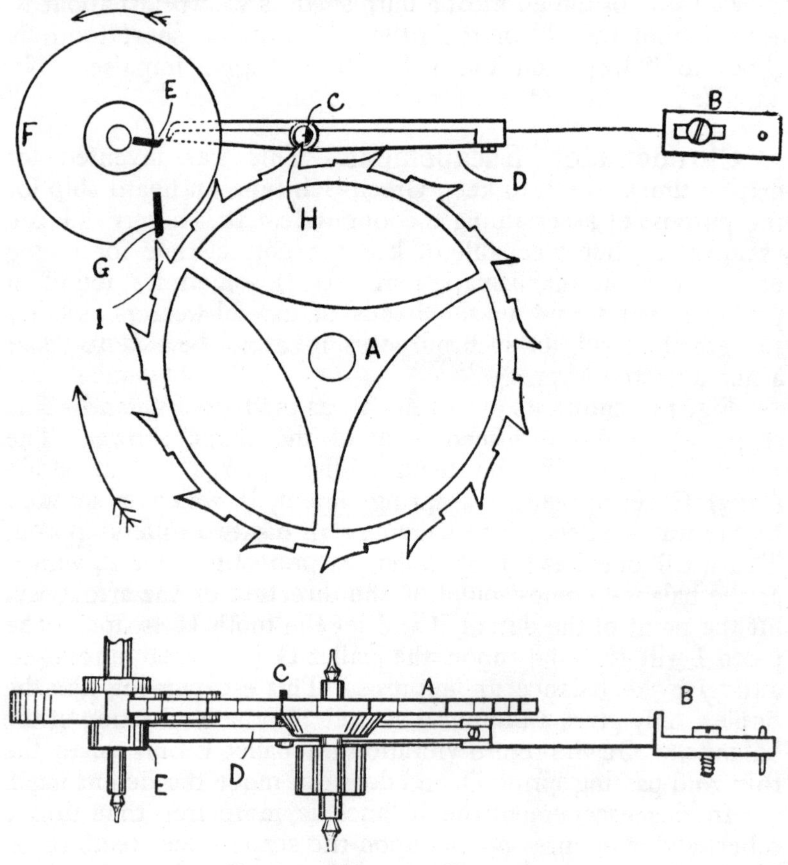

FIG. 145.—Chronometer Escapement.

Broken pallet jewels can be replaced by cementing with shellac. A detent locking stone is half round, as at A (Fig. 146), and should be fitted into its pipe by filing up a half-round brass pin, B, to fill up the interior. This pin should not be wedged in tight, or the jewel will break, but should be an easy fit and cemented altogether with shellac. Warmed pliers are used to hold the "pipe" and run the shellac.

A new gold spring may be made from any scrap of gold wire by hammering out and filing up. Some are riveted and others are screwed to the detent. A screwed spring should have an oval screw hole for adjustment. The point should be square, and the edge clean, burnished, and free from burrs.

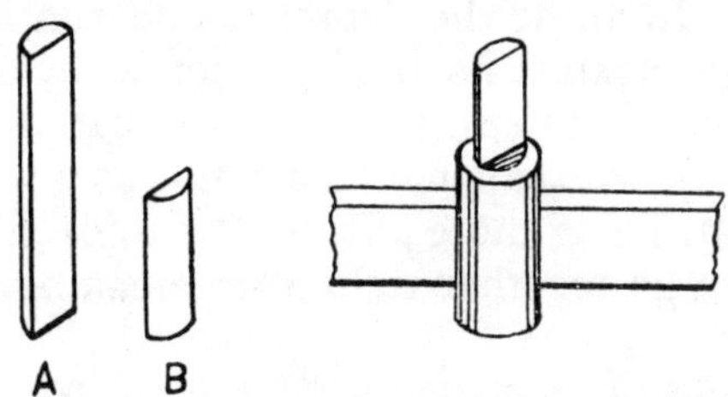

FIG. 146.—Fitting a Detent Locking Stone.

The gold spring should not be of equal thickness throughout, but the point may be thickest and the part near to its fixed end the thinnest, so that nearly all its bending is done near to where it is fixed.

A detent is a delicate thing to make, perhaps the most delicate part found in watchwork. The greatest care must be used in handling one. It is filed up from the solid, foot, spring, and body all in one piece. In making a new one, a length of steel is taken, the position of the foot screw-hole marked, drilled, and filed oval for adjustment, then the hole for the locking stone drilled and a brass pin fitted for trial. It is then roughly shaped up, hardened and tempered, the spring portion being left fairly thick until last. It is now worked down to size all over, and its extreme point softened and turned round for the gold spring to lie upon. When adjusted for length accurately so as to lock the scape wheel in the exact position, the steady pin hole is drilled in its foot and the spring thinned, first by filing, then by polishing down with oilstone dust and oil on a flat polisher until it is thin enough. Finally, it is polished with red-stuff and oil.

Detents are occasionally seen that have broken and been repaired. In nearly every case, though the watch has been made to "tick" by so doing, its timekeeping qualities have suffered badly. In fact, it is no longer an instrument of precision. The only remedy for a broken or damaged detent is to make a new one.

In putting this escapement together, care should be taken to see that the edge of the gold spring does not touch the scape wheel or the banking screw, and that the detent point or the gold spring point does not touch the flat of the large roller.

Some foreign chronometers have a pivoted instead of a spring detent. In these the detent pivots run in jewel holes, and it is brought against its banking screw by a spiral spring like a hairspring, or sometimes by a thin flat spring generally made of gold. Such escapements are not so good as those on the English plan, being more affected by thickening of oil, and not so certain to go together right after cleaning.

Other forms of escapement have been made, such as the "Virgule" and others on its plan; also combinations of the lever and chronometer. But they are not common, and if met with, will offer no difficulties to a workman who has mastered the action of the escapements described in this chapter.

CHAPTER XI.

BALANCES AND HAIRSPRINGS, ADJUSTING AND TIMING.

A Balance is a fly-wheel, and may be circular or any other shape; it may run true or not. The one essential is that it must be in perfect poise; that is, it must have no heavy part, and when rested on the straight edges of a poising tool it must have no tendency to settle in any one position. It is an advantage to have as much of the weight of a balance as possible in its outer rim. It is also an advantage, principally for the sake of appearance and for convenience, that a balance should be circular and should run fairly true.

Balances may be plain or compensated. Plain balances are made of gold (so as to be non-corrosive and keep clean), brass, or steel. They generally have three light arms and a true circular rim.

Temperature Error.—A balance is controlled by its attached hairspring, and the time of its vibrations depends upon the strength of the hairspring. A strong spring will cause rapid vibration, and a weak spring a slow motion. Both hairspring and balance are affected by changes in the temperature. The balance expands as it is warmed, and contracts as it is cooled. When it expands the arms lengthen and the rim increases in size, removing the weight further from its centre and causing it to move more slowly. When a spring is warmed it loses some of its force, also causing the balance to move more slowly. The combined effect of a change of temperature of 45°Fahr. (from the cold of a bedroom dressing-table to a warm waistcoat pocket, say from 40° to 85°) is to cause an uncompensated watch to lose several minutes per day. Temperature also affects the depths a little by causing the wheels to expand; it alters the strength of the mainspring, and affects the fluidity of the oil. The combined effect of all these is the "temperature error" of the particular watch in question, and varies in each individual watch. Therefore machine-made watches cannot be turned out "compensated," but each one must be finally adjusted by itself.

Compensation Balance.—The net result of a change of temperature is to cause a serious loss in heat and a gain in cold. To counteract this the compensation balance is used. Fig. 147

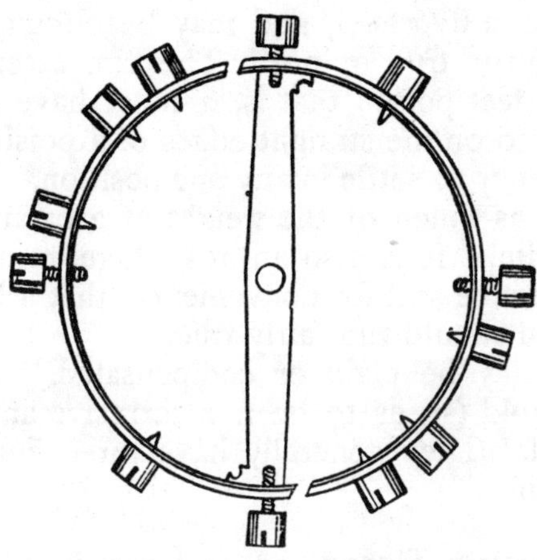

FIG. 147.—Compensation Balance.

shows the construction of an ordinary compensation balance. It is a circular rim with a steel crossbar. The rim is bi-metallic, being steel inside and brass outside, and cut through at two opposite points. Brass is more affected by heat than steel, and in a rise of the temperature the outer brass will lengthen more than the inside steel. The effect of this is to curve each half of the rim inwards and bring the weight of the balance as a whole nearer to its centre. This causes the watch to go faster, and, if the amount of the inward movement of the rim is exactly sufficient, will compensate the tendency of the watch to lose. These balances are weighted by screws fitted in a series of tapped holes all round the rim. By moving the screws nearer to the free ends of the two segments of the rim the effect of the balance is increased; by moving them towards the fixed portions the effect is diminished. Therefore adjusting for temperature consists in trying the watch in cold, then in heat, and moving the screws according to the performance of the watch, until its rate in cold (40° to 50°) is equal to its rate in heat (80° to 90°).

In Fig. 147 there will be noticed four screws at equal distances from each other, with long taps. These are the "quarter screws." They are never moved for temperature adjustment, but are for poising the balance or for small timing alterations. Drawing one out a little makes that part of the rim heavier. Drawing out an opposite pair will slow the watch. Turning a pair in will make it go faster. In the best balances "quarter nuts" are fitted instead of screws. Fig. 148 shows a quarter nut. It is a gold nut, turning on a fixed steel screw, and is not so liable to work loose as a plain quarter screw from frequent turnings. A quarter nut is split, and slightly sprung on to its screw to move firmly and not get loose. To turn these nuts a split screwdriver blade like A (Fig. 148) is used.

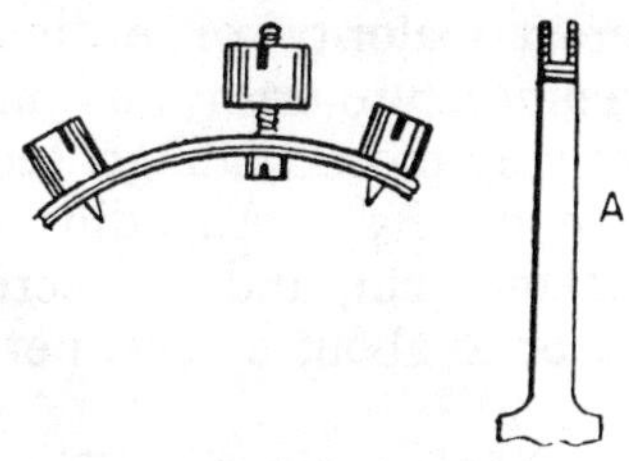

FIG. 148.—Quarter Nut and Screwdriver.

In making a compensation balance, a steel disc is turned up true, with a central hole. It is then covered with molten brass, which adheres to it all over. The brass is filed off the flat sides of the disc and the central hole cleared; then the surplus brass is turned off the edge, leaving the thickness required. It is hammer-hardened and turned smooth. The interior of the balance is cut out by turning and filing, leaving the crossbar; the rim is drilled and tapped in a dividing engine, and finally cut through at two opposite points, after being mounted upon its staff by the escapement maker.

When a balance is cut the unequal hardness of the brass and steel composing its rim generally causes the two segments to go out of truth. They are trued by removing all the screws and putting in the turns to note where they depart from the circle, and bending with the fingers as far as possible or with brass pliers. Absolute truth looks very nice, but is not essential.

Compensation balances that have been running some time generally go a little out, and, so long as they can be poised by means of the quarter screws, are best left alone. The only way to true them is to remove all the screws. Before doing this it is best to make a sketch of the balance rim, noting the position of the screws, so as to replace them as before.

A watch with an uncut compensation balance is no better than one with a plain balance; but if the balance be cut and trued as described it will be greatly improved. A watch with a cut compensation balance, not specially adjusted, like the great majority of ordinary watches, is cured of *most* of its temperature error, and may generally be depended upon not to vary more than 30 secs. from its rate in one day between 40° and 85°. An adjusted watch is one with a compensation balance cut, and the screws arranged by trial to reduce the error to about 2 secs. per day or less.

Hairsprings.—Fig. 149 shows an ordinary flat hairspring.

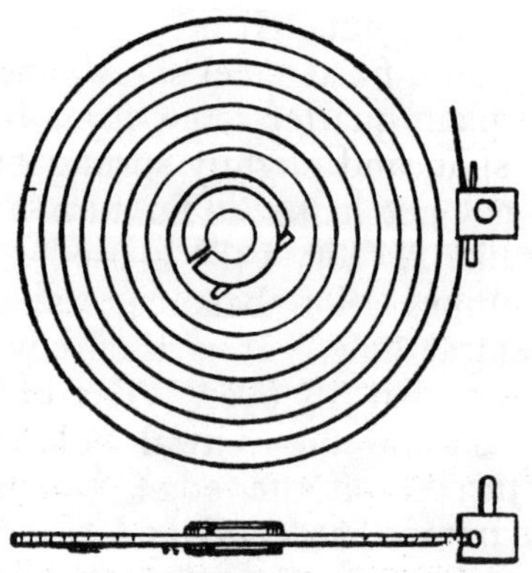

FIG. 149.—Flat Hairspring.

Its outer end is pinned into a fixed stud and its inner end into a collet.

The stud may be a small square of brass fixed in the watch plate, as in some English full plates, a small square of brass pushed friction tight into the balance cock, as in many Genevas, or a steel stud screwed to the plate or the balance cock, as in the best English levers, American and Swiss watches. The collet is in most watches a small brass circle turned to fit the balance staff friction tight, and split to give it a certain amount of spring, as in Fig. 149. A better form of collet, used in the best English levers, is that shown in Fig. 150. It is a circle of steel, hardened and tempered, made to accurately fit the staff. It has two flats filed upon it, making it nearly oblong. Such a collet goes on more truly, and is not liable to become damaged; it also allows the hairspring to be trued in the centre more easily.

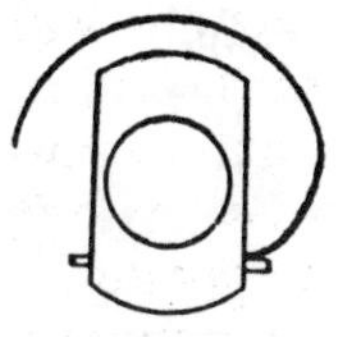
Fig. 150.—Square Steel Collet.

The spring itself is of steel wire, ribbon-shaped, drawn hard, and coiled up into a close spiral. The closeness of the coils depends upon how many are coiled up together. Thus, if three lengths of wire are coiled up together, when released the resulting spirals will be rather open. If two are treated together the coils will be closer. If only one is coiled up on itself the coils will be very close and nearly touch. The closer a spring is coiled the longer it is for a given diameter.

The strength of a spring depends on its thickness and width—that is, its stiffness—and its length. Hairsprings are sold in small packets, numbered according to their strength in a series of numbers which differ with each maker, and are only useful to compare one spring with another of the same make. It is, therefore, best in hairsprings to keep to one make only for ordinary quality springs, and to send specially for one of extra good quality when required.

To select a new hairspring, the number of beats required in an hour must first be known. This is termed the "train" of the watch. Most watches have 18,000, 16,200, or 14,400 trains; that is, they make that number of beats in one hour. All Genevas and American watches have 18,000 trains. Old English watches often have 14,400 trains, and many more recent ones 16,200, or 15,400. But it is probable that all watches in the future will be made 18,000.

The train of a watch can be ascertained by multiplying the numbers of teeth in the centre, third, fourth, and scape wheels, and dividing the result by the third, fourth, and scape pinions. Twice this number is the train.

The following table gives the usual trains found, with the number of beats in one hour, in one minute, and in 20 seconds.

Description of Watch.	Centre wheel.	Third wheel.	Fourth wheel.	Scape wheel.	Third pinion.	Fourth pinion.	Scape pinion	Train.	Beats per min.	Beats in 20 secs.
Geneva or American	80	75	70	15	10	10	7	18,000	300	100
,,	80	75	80	15	10	10	8	18,000	300	100
,,	64	60	60	15	8	8	6	18,000	300	100
English	80	75	80	15	10	10	8	18,000	300	100
,,	64	60	70	15	8	8	7	18,000	300	100
,,	64	60	63	15	8	8	7	16,200	270	90
,,	64	60	60	15	8	8	7	15,400	257	86
,,	64	60	56	15	8	8	7	14,400	240	80

It will be noticed that in all these trains the fourth wheel makes one revolution per minute; and when the fourth wheel has ten times as many teeth as the scape pinion, the train is 18,000; when nine times, the train is 16,200; and when eight times, the train is 14,400.

In counting vibrations when fitting hairsprings, double vibrations only are counted, thus each time the balance comes to the left is counted one. In this way an 18,000 train counts 75 in half a minute; a 16,200, 67; and a 14,400, 60; or in proportion for twenty seconds or a full minute.

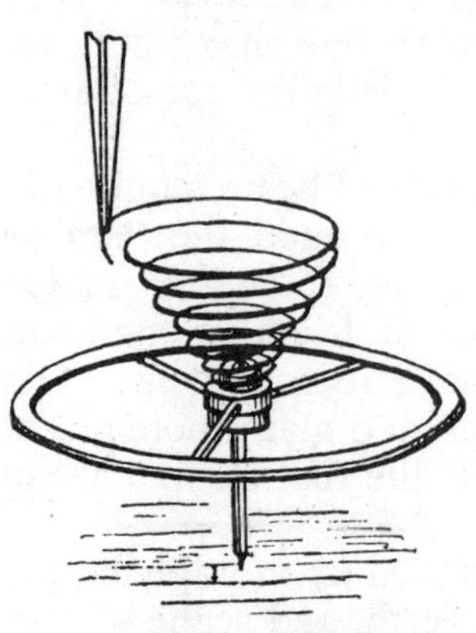

FIG. 151.—Counting the Vibrations of a Balance.

Pick out a spring that is a little too large in diameter to lie in the curb pins of the index. Lay the spring on the balance and press the collet down upon it to temporarily hold it in place. Hold the end in tweezers and let the balance hang down with its lower pivot resting on a watch glass. With a turn of the fingers and tweezers, set the balance vibrating about half a turn each way, or more. Then hold it perfectly still and steady and it will continue to vibrate for a full minute. The vibrations can be counted for twenty seconds to see if it is anywhere near what is required. Fig. 151 shows how the spring and balance are held. If the balance moves too slowly, select a stronger spring, if too fast, a weaker one, and try again. The spring should be held in the tweezers at the exact point at which it will have to be pinned in its stud; thus, if the spring is two coils too large to go in the curb pins, it must be counted while held two coils from the end.

Finally, select one that counts one double beat slow in a full minute. Lay this on a convex watch glass on white paper, and with tweezers and a needle point, or two pairs of tweezers, break out short pieces (about $\frac{1}{4}$ turn) at a time from the "eye" or inner coil, gradually enlarging the central opening until it will go easily over the collet with a little room to spare. Then bend a short piece sharply inwards to pin in the collet, as in Fig. 152. For breaking out and bending springs, a needle set in a handle, filed up, and slotted as shown at A (Fig. 152) is useful.

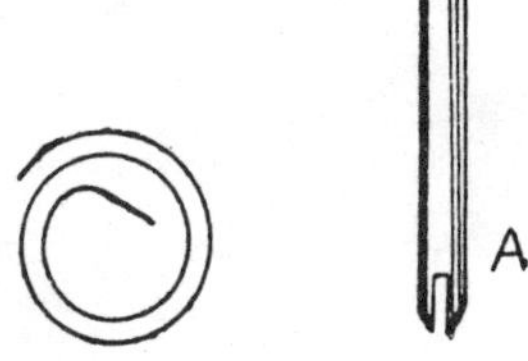

FIG. 152.—Eye of Hairspring.

To pin the spring in its collet, put the collet on a broach, pass the spring over the broach, and insert the end with tweezers. File up a fine burnished brass pin, and file a flat upon it, making it D-shaped. Insert this pin to see if it fits, with the flat against the spring. Cut off any that projects, and try again. Cut off the end until the pin goes in half-way through the hole. Then lay the pin on the filing block (still in the pin vice), and half cut it through with a knife; insert it finally, and break it off in the hole. With very strong tweezers press the pin well home. A hairspring must be pinned tight in its collet, or nothing can be done with it. When pinned in, put the collet and spring on a turning arbor in the turns and revolve with a light bow. Set it flat with tweezers so that it runs true. Note if it is true and concentric in the "eye," and in which direction it is out. Take it out and lay on the watch glass to bend true as required, try in the turns again, and so on until it is both flat and true. Put it on the balance and count for a full minute. If correct, break off the outer waste coils and pin it in the stud with a flatted pin. To set the spring central and flat on the balance cock or plate, pin it in as in Fig. 153, and bend it until the outer coil lies between the curb pins without strain, the spring stands level, and the collet is central with the jewel hole. If all these points are attended to before the spring is put in the watch at all, there will be no bending or cramping needed afterwards, and the balance and spring will go straight in, lie true and flat, be central, and free of everything.

The watch can be started and set by the seconds hand with the regulator clock. A loss or gain will be quickly seen.

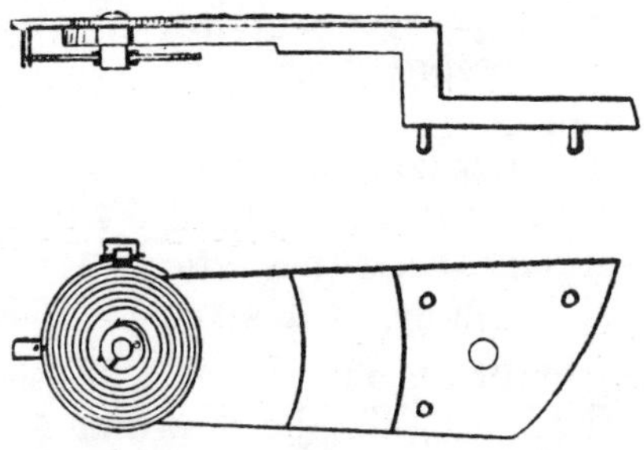

FIG. 153.—Setting a Spring true and flat.

If too slow, take the spring up a little and re-pin in the stud; if too fast, let it out. It is here that the advantage of counting the spring one beat slow is found. If the balance is a compensation, or a sham one with screws, to make it go faster, a pair of screws can be reduced by filing or drawn out altogether. If too fast, a pair of screws can be added, or for a small alteration, "timing washers" can be added under the screw-heads. These are small washers stamped from thin brass sheet, and can be bought by the gross.

Balance screws can be reduced by holding in a small pin vice of the pattern shown in Fig. 154, and revolving it in the

FIG. 154.—Small Pin Vice.

fingers as the file is passed across the screw-head. Smooth them with a $\frac{3}{0}$ emery buff and polish with a rouge-on-leather buff. Always be careful to leave the balance in perfect poise after any alteration.

The hairspring and its collet can be removed from the balance staff by levering up with a sharp pocket-knife alternately from either side. A collet can be turned round by inserting the thin blade of an oiler into its slit and using it as a lever.

A flat hairspring should be pinned in at equal turns, as shown in Fig. 149; that is, it should consist of so many complete turns, finishing against the point where it starts from the collet. Such springs time better in good watches.

Sometimes when a flat hairspring is pinned in its collet and in its stud it will not lie flat. It is dome-shaped or cup-shaped, owing to the stud hole not being drilled level with the collet hole. In such a case the innermost coil of the spring, just where it leaves the collet, must be bent up or down as required to get the spring flat, as shown at A (Fig. 155), and the spring then set flat and true in the turns.

FIG. 155.—Setting a Hairspring.

In some very flat Geneva watches and undersprung English levers there is very little room for a hairspring without it touching the plate, the balance arms, the stud, or the index. For these watches a spring made from specially *narrow* wire must be got from the hairspring makers.

To lower a collet and spring bodily, or to lower the top of a collet to prevent it fouling the balance cock, the collet, together with the spring, may be placed on an arbor and turned down. A very sharp graver must be used and light cuts.

Breguet Springs.—Fig. 153 shows a breguet or overcoil spring at A. This is perhaps the best form of hairspring for any watch. Sometimes a double overcoil is made, as at B, but apparently has no advantage. Helical or cylindrical springs, as at C, are used in marine chronometers and some pocket watches, but do not seem to perform any better than a single overcoil breguet.

Breguet springs are made from flat springs by the workman who springs the watch. For this purpose hardened and tempered hairsprings are best. These can be obtained by sending the balance to the spring maker and telling him the train. Sometimes springs of palladium wire are used, especially in "non-magnetic" watches. This metal does not rust, and cannot be magnetized. Balances also are made with palladium in place of steel for the same purpose.

Having procured or selected a spring, preferably a close coiled one, with a diameter equal to half that of the balance rim, proceed to put it on like a flat spring, getting it *perfectly*

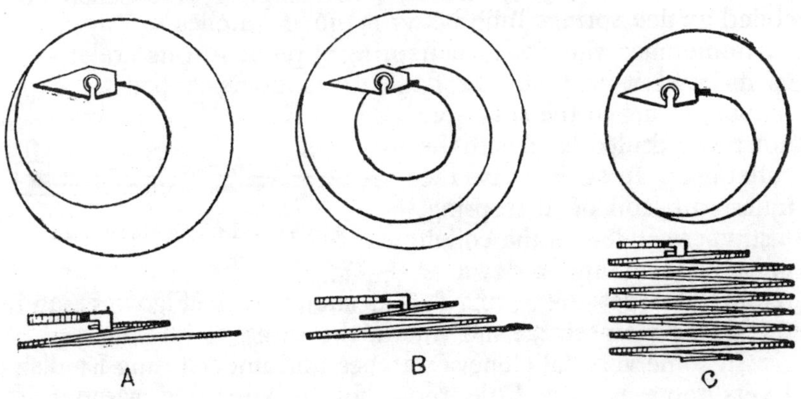

FIG. 156.—Forms of Hairsprings.

true in the eye and flat. Count it exactly to time, and break off all surplus coils.

Unless a breguet spring is quite true in the eye it looks very bad and will not time well. When a spring is true there will be seen one coil, about midway between the outer coil and the eye, that apparently stands still. This is, of course, only an optical illusion. If a spiral be revolved in one direction, the coils all appear to run outwards; in the other

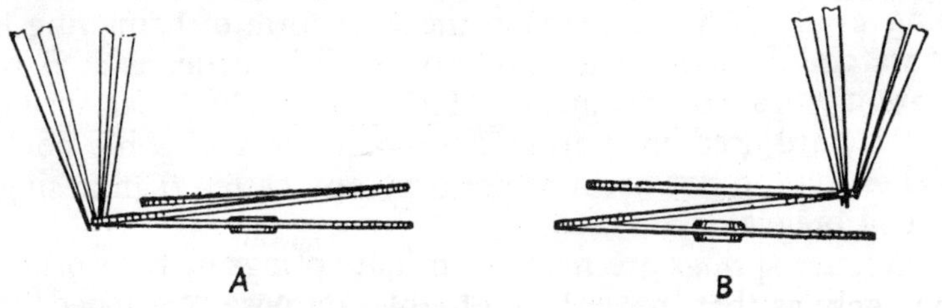

FIG. 157.—Bending up an Overcoil.

direction they will all seem to go in. The "stationary coil" is merely that point at which this optical effect is exactly neutralized by the mechanical opening and closing of the spring caused by the vibration of the balance and which takes place in the opposite direction to the optical effect. However, this coil should be seen in a true spring, and should apparently lie motionless. If it jumps or shakes, the eye of the spring is not quite true.

When broken down to size, take two pairs of tweezers and bend up the outer coil like Fig. 157, A. Hold the spring firmly with one pair and twist the outer coil upwards with the other pair, in a gradually ascending slant. Halfway round the raised coil it will want another twist up, to make the raised part level, as at B. Take a curved pair of tweezers, like Fig. 158, and proceed to curve the raised coil inwards by nipping it up tight. Complete the overcoil with two pairs of ordinary tweezers, to the form shown in Fig. 156, at A. See that the overcoil is quite free from the second coil at the point where it commences to curve inwards. To judge the height required for the overcoil, put the balance in the watch and see how far up the balance staff the level of the pin hole in the stud comes, by sighting it across. Raise the overcoil to this point. If the watch has an index, the last quarter turn of the overcoil must be circular, and of the same radius as the curb pins.

Fig. 158.—Curved Tweezers.

This can be measured with the millimetre gauge. All manipulation of the spring should be done while on a watch glass over white paper. Finally, pin the spring in its stud with a flatted pin, and drive it home hard. With a breguet spring, all timing is done by means of the balance screws, and the spring itself is not disturbed. Set it flat and true as it lies on the cock, so that it stands level and with the collet central over the jewel hole. Sight this across in two directions, as in Fig. 159. Then put it in the watch with the balance and see if it is flat and correct. If the spring is hollow the overcoil is not high enough, and must be raised a little; if domed it is too high.

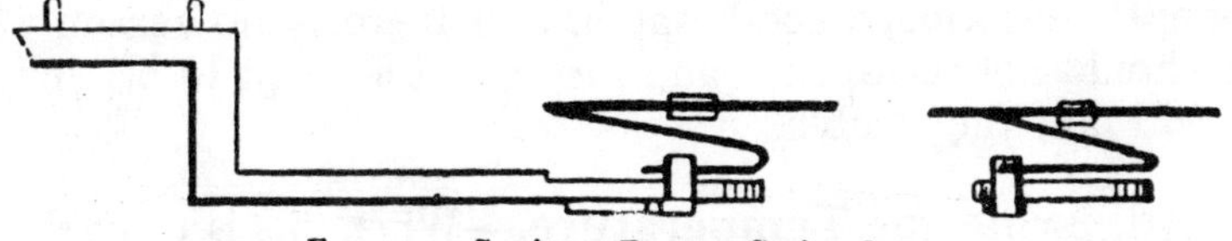

Fig. 159.—Setting a Breguet Spring flat.

A great deal has been said and written about breguet springs being pinned in at equal turns, like Fig. 160, so that the spring starts at the eye at the same point or in the same direction as it is pinned in the stud. There appears to be no real advantage in this, as breguet springs, unlike flat ones, may be pinned in anywhere, simply at haphazard, and act just as truly and nicely as at equal turns. Watches so sprung by the writer have many times obtained 80 marks and over at Kew, and taken high positions in Greenwich Admiralty trials.

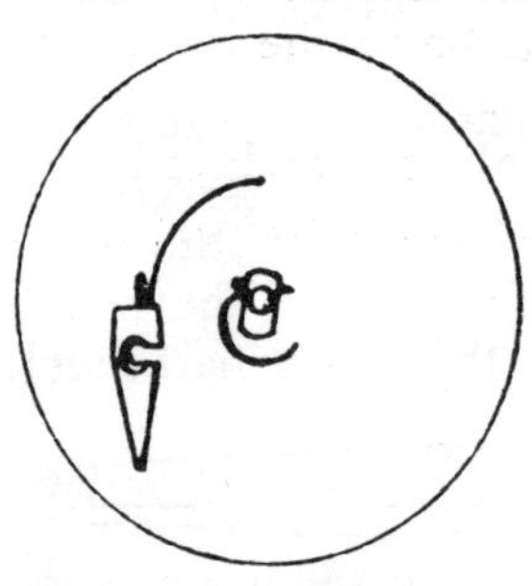

FIG. 160.—Breguet Spring pinned in at Equal Turns.

Time the watch by altering the weight of the screws. Additional weight can sometimes be given by changing a pair of brass screws for a pair of gold ones, by changing common gold for good gold, or gold for platinum.

If the watch in hand is a good one, with a compensation balance, it may be desired to adjust it for temperature and positions. At all events, one thing is quite certain; if the watch ever had been so adjusted, by the very fact of re-springing it all adjustment is gone. A new hairspring will require a fresh adjustment for temperature to suit itself, and disturbing the balance screws altogether upsets the position rates.

Good watches often require re-springing. Sometimes a little rust appears on the coils. This is fatal, as it eats further and further in, causing the watch to lose at a gradually increasing rate. Sometimes they get damaged by a careless workman or wearer. From one cause and another a repairer often is called upon to re-spring a good watch, and there is no reason why he should not re-adjust it and turn it out a credit to himself, if he will take the trouble.

Adjusting for Temperature.—When this is to be done, some sort of oven is necessary to keep the watch at about 80° to 90°. The cold can generally be managed in England. In the winter 40° to 50° is easily obtained in a workshop, too easily in fact; while our summers are not so sultry that a few days cannot be found when 55° or 60° can be got without waiting long. With care these temperatures will serve. The oven may be a tin box set on an iron plate, and warmed underneath by being placed on brackets against a wall, about a foot, or less, over a small gas-jet. Or there are many ways that suggest themselves and are suitable to the special conditions of the workshop. A thermometer should be in the oven with the watch.

The watch, when running on time for about two days in the cold, is set by the shop regulator, and its rate noted in a rate book. It is then put in the oven for 24 hours and a comparison made. If 20 seconds slow in heat, it is under-compensated, and one or more screws should be moved several holes nearer to the free ends of the segments. If fast in heat, it is over-compensated, and screws must be moved back. Proceed thus until the watch shows an equal rate, even when tried three days in succession in cold and in the oven. It will be found that moving a pair of screws one hole will make a difference of about 2 seconds per day.

Timing in Positions.—When correct, poise the balance, and again bring to time within about 20 seconds per day. The next thing to do is to get the long and short arcs equal. When a watch is lying down, the balance spins on one pivot and makes a large vibration, say 1½ turns. When it is placed vertically, with 12, 9, or 3 up, the balance runs on the sides of two pivots, and there is more friction. It therefore makes, say, only 1¼ turns.

These are, therefore, the long and short arcs, and in all probability the short arcs are slower than the long ones. This is a fault of the hairspring, and the overcoil can be so shaped that the spring will cause the long and short arcs to be made in equal times.

To test the watch, note its rate lying for 24 hours (do not be tempted to make trials shorter than this, as they are misleading). Then 9 up and 3 up for 24 hours each. These two opposite quarters are tried so as to eliminate the errors caused by want of perfect poise. Suppose the rates shown are as follows :—

Watch lying	+ 3	secs.	per day.
„ 9 up	+ 11	„	„
„ 3 up	− 5	„	„

Here the rate 9 up is 8 seconds faster than lying, and 3 up

8 seconds slower. The mean of the two is equal to the lying rate, showing that the long and short arcs are equal, and the difference between the two quarter positions is only a question of poising the balance by a touch of a quarter screw.

Suppose the rate is as under :—

Watch lying + 3 secs. per day.
,, 9 up − 2 ,, ,,
,, 3 up − 13 ,, ,,

This is a more likely rate, and shows that 9 up the watch is 5 seconds slow, and 3 up 16 seconds slow. The mean is $10\frac{1}{2}$ seconds, which is the amount the short arcs are slow.

To correct this, the form of the overcoil is slightly altered to make it more flexible. Therefore bend the overcoil to make the curve more symmetrical, an easy flow from start to

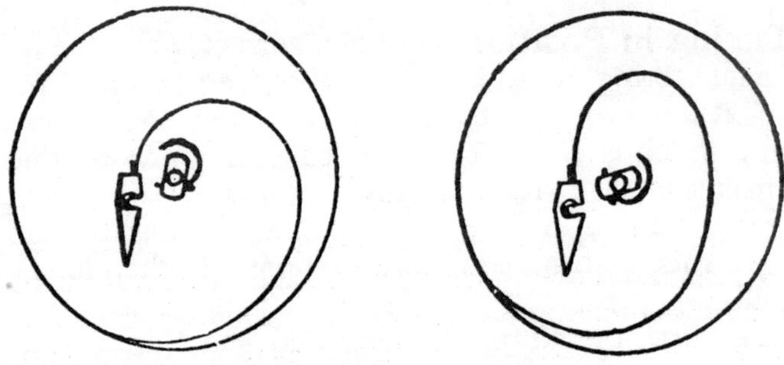

Fig. 161.—Forms of Overcoil.

finish, with no angles or straights, is the most flexible, and will quicken the short arcs. Fig. 161 shows the overcoils of two watches that have taken Kew "A" certificates with over 80 marks, and may serve as a guide. If this does not do it sufficiently, cause the spring to open or develop more in the direction opposite to that in which it is pinned in its collet. This can be easily done by a trifling bend, quite unnoticeable, in the overcoil.

It is at this point that there is sometimes a slight advantage in having the spring pinned in at equal turns. When the overcoil is a short one, it is not so easy to cause the spring to develop in the required direction unless it is so pinned in.

With a longer overcoil or a double overcoil, this difficulty disappears.

When the mean of the short arcs equals the lying rate, correct the positions by trying 12 up, 9 up, and 3 up. A loss in any position shows that the upper part of the balance in that position is heavy. Put that quarter screw in a trifle, or draw the opposite one out. In this way the position rates can be got equal all round, unless there are escapement faults which cause errors of their own.

Such faults are, an imperfectly poised lever, worn or bent pivots, unequal endshakes of scape wheel and pallets, oval jewel holes, wide pivot holes, unequal banking shake of the lever, or sloping bent banking pins and guard pin.

Also a breguet hairspring should have no play between the curb pins. It should touch each pin, but not be nipped between them or strained by them. Watches with no index, "free sprung," time much better than those with curb pins. One troublesome source of errors is absent in them; but a free-sprung watch that is not perfectly adjusted for temperature and in positions is a nuisance.

The method of position timing just explained is condemned by some writers as bad. They say that alterations of the quarter screws upset the poise of the balance. This is not so. The poising tool at its best is imperfect. There is a small error always present. Also the balance as poised on the tool has not the hairspring and collet on. This disturbs the poise in itself, seeing that a part of the weight of the hairspring is borne by the stud and a part by the balance, and there is no means of ascertaining how much or in what manner. The balance is poised as nearly as possible and then put in the watch. There its poising is perfected by the only possible means, viz. noting its rates in the various positions and touching the quarter screws accordingly. It has been recommended that the collet be put on the balance when poising, with its pin and a small piece of hairspring inserted as a refinement. This is absurd, as when subsequently sprung the poise is altogether upset many times by altering the balance screws to bring it to time and adjust for temperature.

A theoretical balance is one having *all* its weight in its rim, no friction at its pivots, and not connected in any way with an escapement. For such a balance, a mathematician can plan an overcoil for the hairspring that will make the long and short arcs equal. A number of these curves have been planned and published, and many watchmakers have copied them in the belief that they solve the difficulty. But as the theoretical balance is impossible in practice, so the theoretical curve is inapplicable. Just as friction at the pivots varies, balances have long and short staffs, and escapements differ, so the curve to obtain isochronism differs in each watch, and must be made to suit each individual case by trial.

Variations between the positions of dial up and dial down are caused by some parts of the escapement having unequal endshakes, unequal sized pivots, bent banking pins or guard pin, or it may be a little dirt in a jewel hole.

Fig. 156, at A and B, and Fig. 160 show overcoils made for an index in which the first part of the curve is a portion of a circle. Fig. 161 shows two overcoils for watches with no index—"free sprung," as they are termed. In these no portion of the curve is necessarily circular.

If a very exact temperature adjustment is wanted, the watch should be first adjusted in heat and cold until within three or four seconds per day; then the isochronism—equal times of long and short arcs—got correct. It may then be run several months to settle. All balances go off their temperature adjustment a little as the two metals composing the rim settle to one another. It is then readjusted for temperature, the balance finally poised, and the positions got right last of all.

It is a help to a new balance to settle quickly to warm a brass plate and lay the balance on it, to close up the rim; then to cool it by laying on very cold steel. Repeat the process half a dozen times.

Non-magnetic springs are soft and heavy. They are not so nice to handle, being liable to damage and to being bent out of shape. Their weight makes them shake about in the watch during pocket wear and foul the balance arms, the cock, the curb pins, or the stud. This affects the timekeeping. But they are very necessary in electricians' watches. Non-magnetic balances also are soft and easily bent out of truth. They often go out of truth with the mere lapse of time, and altogether upset the rate of the watch. Like the springs, they are a necessary evil.

Some Swiss watches have latterly been made with balances

and springs of an alloy of nickel steel, that is claimed to be hardly affected at all by changes in temperature; but for exact compensation they cannot yet touch the steel spring and compensation balance.

Very cheap watches with plain balances and springs of this alloy have been made, and in their class are a great advance on the same watch with a plain balance of brass and a steel spring. One use to which this alloy has been put in watch-work is to make hairsprings that require very little compensating. It is found that an ordinary balance compensates too much for them if cut as usual; but if cut at the centres of the rim, leaving four short bi-metallic arms, a stronger and stiffer balance is produced, that at the same time compensates quite enough to correct all the errors. This pattern of balance is found now on many Swiss levers.

"Invar" at one time seemed likely to effect a revolution in both watches and clocks. Pendulum rods for clocks made of it have been tested at Kew, and found to expand only $\frac{1}{20}$ in. per mile per degree Centigrade, which is, of course, quite an inappreciable quantity; but in watches the applications of invar are still in an experimental stage, and that they will be equally far-reaching in the near future seems doubtful. Compensation balances made of invar and brass instead of steel and brass have not proved so satisfactory as was anticipated. Possibly the difference between the expansion of the two metals is too great to be stable.

Watches in which the balance comes at or about the centre of length of the balance staff have been noticed to time better and go more steadily than those in which the balance is at one end of the staff. The reason of this is not very apparent, but it is none the less true. A peculiarity the writer has found in Karrusel watches (in which the balance comes in the centre of the staff and the staff is extremely short) is that, when first sprung and tested for isochronism, the short arcs are found *fast*, whereas in most watches they are invariably more or less *slow* to begin with. An explanation of this is also wanting.

A good watch wanting a breguet spring is worth fitting with one fire hardened and tempered, although these are rather expensive, and cost from 1*s*. 6*d*. to 3*s*. 6*d*. each. For ordinary fair quality work, the usual hard drawn springs, costing about 4*d*., are good enough.

Repairing Hairsprings.—Geneva watches are often badly treated by their wearers. If they stop from any cause, ladies often stir them up with a pin; the inevitable result being a bent or tangled up hairspring. To get such a spring fairly straight again is a task requiring some skill and patience. Take such a spring off the balance and lay it on a glass. Begin at the centre and follow it round, coil by coil, until the first bend or departure from truth is detected. With tweezers and needle-point correct this. Follow round further and correct, until the end is reached. Then proceed to get it flat. Do this in the same way. Begin at the centre; hold the spring up edgeways to the light, and note where it first departs from the flat. With two pairs of tweezers, correct it as in Fig. 157, and proceed to the next point, and so on. Generally speaking, springs that look all tangled up and done for will, on close inspection, be found to have one or two sharp bends, which, when corrected, bring the spring right again as if by magic. Finally, place the collet on a turning arbor and set the spring flat in the turns; affix the stud to the cock or plate, and set it flat in the watch (without the balance), as in Figs. 153 and 159, and see that it passes between the curb pins properly, and that the collet is central with the balance holes.

A spring that has been pulled up like A, Fig. 162, may be flattened by putting the collet on a broach and pulling the

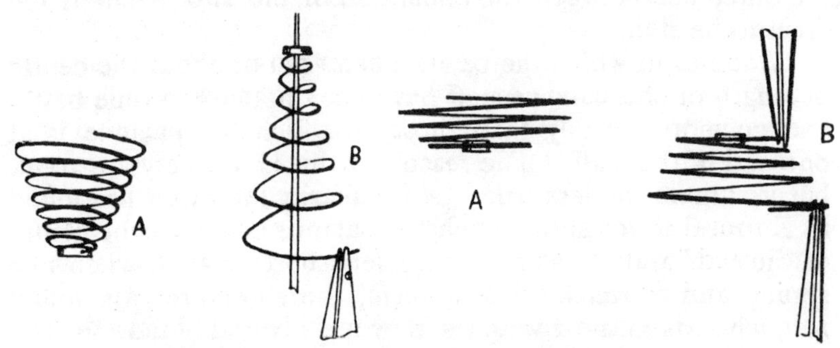

FIGS. 162 and 163.—Flattening the Coils of Bent Hairsprings.

stud down with tweezers in the reverse direction, as at B. One that is partly pulled up, as at A, Fig. 163, may be treated with two pairs of tweezers, as at B, and sprung flat.

Occasionally in brushing a balance the hairspring gets

tangled up, it may be round the balance rim or round its stud. If round the balance rim, lever up the collet and get it off; then turn the spring round and round until it screws off the balance rim. If round its own stud, unpin from the stud, and with a needle-point begin at the centre coil and run the needle round the spiral until the outer end is reached. This will disentangle it safely.

In many watches a shake during wear will frequently jolt the hairspring out of the curb pins, or shake the second coil over or into them. To avoid this, the curb pins should be as long as possible; or in some cases they may be bridged over by arching them at the points and bending towards each other till they touch, making a complete loop enclosing the outer coil. To replace curb pins, file the old ones off flush with the under side of the index, and lay it over a hole in a steel stake, punch them through with a needle, the point of which has been flatted on an oilstone. New curb pins should be filed up nearly straight and well burnished.

In common watches the hairspring usually has some play between the curb pins. This is a source of error, and makes the watch go slower in the hanging position than when lying down. A good watch should be allowed no play here. Each curb pin should just touch the outer coil of the spring lightly without nipping it between them.

Many old watches had plain uncompensated balances and "compensation curbs." A compensation curb was a bi-metallic strip of metal, made so that its movements caused by heat and cold opened and closed the curb pins, thus keeping the watch to time. This was the earliest application of the bi-metallic principle to the compensation of watches, and has been completely superseded by the compensation balance.

From the foregoing, it will be readily understood that an "adjusted" watch takes a long time to get perfect; and its good performance depends upon the fine adjustments of the balance spring and the poise of the balance. When such a watch has an accident and requires a new balance staff, its position rates are at once gone. Several weeks' careful rating are required to re-adjust the poise of the balance.

Careless handling of the balance or hairspring will destroy all the adjustment the watch ever had. Such watches should therefore always receive the greatest care.

CHAPTER XII.

MOTION WORK, HANDS, AND DIALS.

Motion Work.—The ordinary motion work (Fig. 164) consists of a steel cannon pinion, A, so-called from its shape, a minute wheel, B, and pinion, C, generally of brass, and a brass hour wheel, D, revolving upon the pipe of the cannon pinion. The teeth of these wheels are so proportioned that the hour wheel makes one revolution to twelve of the cannon pinion.

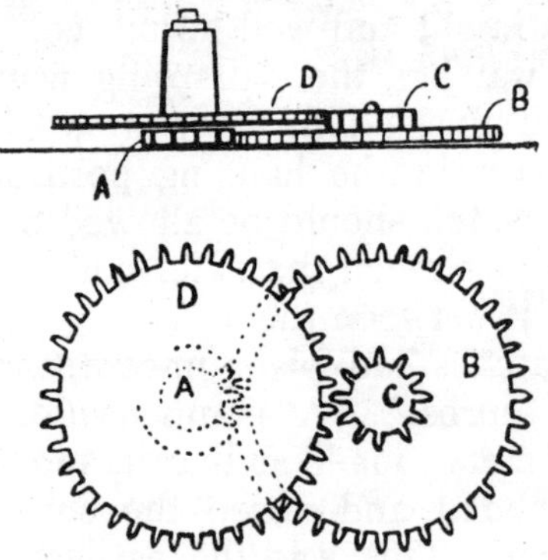

Fig. 164.—Motion Work.

Rough cannon pinions are bought, and require fitting to the centre arbor; the outside has to be turned to fit the hour wheel, and the top cut to length or squared to fit the minute hand. Finally, the "snap" has to be made. Rough hour wheels require broaching out to fit on the cannon pinion, turning to revolve freely under the dial with a little play, and turning to fit the hour hand. A rough minute wheel and pinion has only to be broached out to fit upon its stud, and the pinion reduced in height to be quite free of the dial.

Motion wheels are often lost, and to some workmen it is

more or less a puzzle to find out the right number of teeth for the new wheel.

The usual trains found in key-wind watches and keyless Genevas are as under :—

Cannon pinion.	Minute wheel.	Minute pinion.	Hour wheel.
10	30	8	32
8	32	10	30
10	40	12	36
12	48	14	42

Modern keyless English levers generally have cannon pinion 14, minute wheel 28, minute pinion 8, hour wheel 48; and the minute wheel is of steel for strength, as the hands are set through it.

Any other train can be calculated, remembering that the minute wheel and hour wheel multiplied together and divided by the cannon pinion and minute pinion must = 12.

An English centre seconds motion train is generally cannon pinion and minute wheel 28, minute pinion 6, hour wheel 72, or minute pinion 7, and hour wheel 84.

Tightening the Hands.—A cannon pinion is often loose upon its arbor and turns too easily. Some are snapped on by having a hollow filed in one side and a centre-punch mark in its thinnest place. This makes a bulged place inside, which springs into a slight groove turned in the centre arbor. To tighten one of these, punch the snap, resting the cannon pinion on boxwood. Some have circular snaps. A groove is turned round the cannon pinion and the thin bottom is burnished in. To tighten one of these, hold it in a split chuck in the lathe and reburnish the bottom of the groove. If these methods fail, the arbor may have some rings made round it with cutting nippers. Or if the pinion has a square for the minute hand to fit on, two opposite sides of the square may have a centre-punch mark deeply made on the outside. A cannon pinion that does not turn upon its arbor when the hands are set may be tightened by inserting a hair from the watch brush and pushing the pinion on tight.

A loose set-hands arbor, going through a hollow centre pinion, may be tightened by ringing it with cutting nippers, or by rolling it between two files under pressure. Never hammer them to tighten; it only knocks them out of truth.

An hour wheel that rocks badly upon the cannon pinion can have a new pipe put in very easily by means of a watch lathe, holding the wheel in a step chuck and turning the pipe clean out. A new pipe is turned from clock stopping wire and riveted in.

An hour wheel that has too much shake under the dial may have one or two paper washers placed over it, or a thin washer made of metal foil curled up at the edges to form a light spring and keep the wheel down.

The motion work in some cheap watches is very roughly made, and the teeth have burrs left on them, making the depths tight. Such wheels may be brushed through the teeth with a brass wire brush.

Fitting Hands.—The hands are more important than many people think, and badly fitted or faulty hands probably stop more watches than anything else. The hour hand should push tightly on to the hour-wheel pipe. If the wheel pipe is thin, pushing on the hand sometimes tightens the wheel upon the cannon pinion. The remedy for this is to take the hour wheel out of the watch and put the hand on it, then to just turn a broach round inside to ease it. Some hour hands have thick sockets that rub against the dial. These require thinning down in the turns. To open out an hour hand that will not quite go on the hour-wheel pipe, hold it in a pair of hand tongs, like Fig. 104, p. 91, and file it out with a rat-tail file. Do not use a broach. Or it is sometimes advisable to turn the hour-wheel pipe to fit the hand. For this purpose, put it on an arbor in the turns.

A minute hand that fits on a square cannon pinion requires filing out to fit the pinion. Hold it in the hand tongs, and use a square file that cuts on all four sides. Do not attempt to file the *sides* of the square hole, but file its *angles* one stroke in each, round and round the hole. This keeps it square. A minute hand that has a round hole should, if large enough to admit of it, be filed out, and not broached, as broaching leaves a burr. A minute hand centre boss that is too thick should not be filed, but always reduced by turning. This keeps all surfaces true and flat.

A seconds hand requires broaching to fit on its pivot. If the pipe is too long, shorten it by laying it upon a piece of boxwood, pipe upwards, and placing a piece of thin brass over it so that the pipe comes through a small hole in the brass and projects upwards. File off the projecting part. To remove the burr, put the hand on a pivot broach and file it off with a pivot file, resting the pipe in a filing groove in the boxwood. A pipe may be reduced in thickness by sticking the hand tightly on a pivot broach and filing the pipe with a pivot file on the filing block, manipulating the broach as if it were a pin vice.

All the hands of a watch should be quite free of the dial, the glass, and each other. The hour hand must always have a little play, showing that it is quite free. A fragment of tissue paper shut under the glass over the minute hand centre will show if the hand touches, or a drop of oil or red-stuff on the hand will mark the glass.

Set-hand Arbors.—A new set-hand arbor in a Geneva watch is fitted by filing in the pin vice.

Rough arbors are bought from the makers with the square formed and the arbor rough turned, the whole being hardened and tempered. Pick out one with a square the same size as the winding square. Hold it by the square in a pin vice, and with a fine file reduce the arbor to a very gradual taper to fit the centre pinion. Very careful filing is necessary towards the end of the operation, and for the final fitting a pivot file should be used. When fitted to the centre pinion a good tight fit, reduce the projecting end to fit the cannon pinion, a good driving-on fit. Then cut off to length, round up, and burnish the end. Reduce the square to length, and file off its four edges a little bevelled, smooth and burnish it in a lathe or screw-head tool.

Some old English $\frac{3}{4}$-plate watches have set-hand arbors made in one piece with the cannon pinion, and the square at the back pinned through. Often the arbors are broken off where drilled through. In such a case it is best to put the cannon pinion in a split chuck, turn the arbor off, turn a true drilling centre, and drill the pinion right through. Then fit a Geneva set-hand arbor of the ordinary pattern.

Some difficulty is sometimes experienced in getting the small pin out of these loose set-squares. If a flatted needle is held in a pin vice and used as a pusher, the square is liable to turn round and cause the needle to scratch the watch plate. To do it safely, hold the movement by the cannon pinion, either by means of a pair of cutting nippers and its square or by holding its body in a pin vice. The pin can then be pushed through firmly with no fear of the arbor turning round.

To polish up such a loose set-square, file a taper steel pin, drive the square on, and smooth and burnish it in a screw-head tool or split chuck in a lathe, holding it by the pin.

Enamel Dials.—White enamel dials are subject to many accidents. A crack can only be left alone. A piece chipped out may be stuck in again neatly with gum, or, better still, "coaguline" or "seccotine" cement. A piece chipped out and lost can be filled in with the special white dial cement sold for the purpose. Run a little of this on warm, like sealing-wax, and when cold file it off flat. Then warm over a flame to just "gloss" it. A loose seconds piece is cemented in with this cement. Scrape the piece clean on its edges, and also the dial edges inside. Hold the dial in pliers and warm it in the spirit-lamp flame, run on the white wax, as you would sealing-wax, and place the seconds piece in position. Hold the dial level, and again warm it until the cement is seen to run well. Finally, file off the surplus cement flat.

A soldered seconds may be re-soldered by scraping the edges clean, applying the acid, and laying the seconds piece in place. Cut off small pieces of solder and lay over the join at short intervals all round. Hold the dial with pliers over the lamp flame until the solder runs. Let cool slowly and wash well.

Any brown discoloration that forms on the dial face by heating may be removed by washing or by a peg point and rubbing.

A dial hole is opened by chamfering a little with a small emery stick (sold on purpose) with a cone-shaped end, and then filing with a rat-tail file and turps. Great care has to be taken not to let the file go so far in as to block, or the dial will crack. The outer edges of a dial may be reduced by filing and turps, or with coarse emery sticks. The back of a dial may be hollowed at one spot to free some of the motion work by rubbing round and round with one of the small circular emery sticks kept moist with water.

Sometimes the feet break off. To replace one, scrape off the back enamel with a graver to lay bare a space about $\frac{3}{16}$ in round. Make a new foot from copper wire, and silver solder it on to a round base $\frac{3}{16}$ in. diameter, as in Fig. 165. Soft solder this on to the dial.

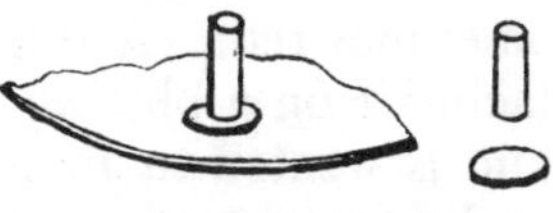

FIG. 165.—Making a New Dial Foot.

The figures of enamel dials are burnt in, and cannot be rubbed off by ordinary means; but names are sometimes added afterwards, in which case they can be removed in a moment by a peg and spirit of wine. A burnt-in name may be removed by using a polishing paste of diamantine and water used on a flat-ended peg. Half an hour's patient rubbing will remove the name and leave a polished surface.

Metal Dials.—Silver dials go a very bad colour in wear. If they have gold raised figures, they may be much improved by washing in soda and hot soap and water and drying in boxwood dust, the gold figures being previously buffed over with a rouge-on-leather buff. This process is apt to remove any painted figures there may be. To restore such a painted dial as new involves bleaching and repainting, which is dial-maker's work.

Gold or silver-gilt dials may be washed in the same way as silver ones. For restoring as new, send them to a dial-maker.

Feet may be put on gold or silver dials by soft soldering. Some gold and silver dials are snapped on by a turned-down edge. Such dials often get loose. To tighten them, lay face down on flat wood with paper between, and burnish the edge a little inwards by hand with a wetted oval burnisher.

To open holes in metal dials, never use a broach; do it all by filing, the dial being held face upwards.

A finger mark may be taken off a new gold dial by clean tissue paper twisted up into a little mop and moistened, using it in the same direction as the engine turned marks on the dial face. Both gold and silver dials, when new, have very delicate surfaces, and should always be handled with tissue paper. The slightest touch of the finger-tip marks them badly. Great care should also be taken in removing the hands not to scratch them, as scratches can only be removed by re-engine-turning the surface.

General Remarks.—Steel hands are sometimes blue and sometimes red. A red hand, however, can always be blued by placing it on a blueing slip and heating. Thus, if a blue hour hand is wanted to match a minute hand, and only a red one can be found to fit, then blue it to match; or if an hour hand is wanted to match a red minute hand, and only a blue one can be found, then blue the minute hand to match it.

An enamelled dial that is thick and prevents the bezel shutting down properly, or that leaves no hand room, may be thinned down on the back with a coarse emery stick until the copper nearly shows bare.

Never leave much white dial cement on the back of an enamel dial round the seconds piece. It breaks away and adheres to the oil round the pivots in the lower plate. Oil dissolves it and forms a sticky compound like glue, effectually sticking the pivots in their holes and stopping the watch.

A seconds piece that is cut out to show the phases of the moon, as in some calendar watches, has a loose centre piece, and a new one can, if necessary, be filed up with a file and turps from the centre of an ordinary seconds dial.

"Up and down" mechanism is sometimes found in good fusee watches to show when the watch is wound up or run down. A small pinion is driven on to the bottom fusee pivot, and works an "up and down wheel," which runs on a fixed stud like a minute wheel. This wheel carries the indicator hand.

Up and down mechanism is occasionally seen in going-barrel watches, but involves great complication and multiplication of parts.

CHAPTER XIII.

CASES.

WATCH cases may be made of gold, silver, German silver, brass, or oxidized steel. Gold keeps the cleanest and, on the whole, wears best. But the material of the case is not of much importance as regards the timekeeping of the watch. The essential is that the case must be sound enough to resist outside pressure, and the covers and bezel must snap tight and fit well.

Dust and dirt will find their way through an astonishingly small space, and very accurate fitting is required to keep a

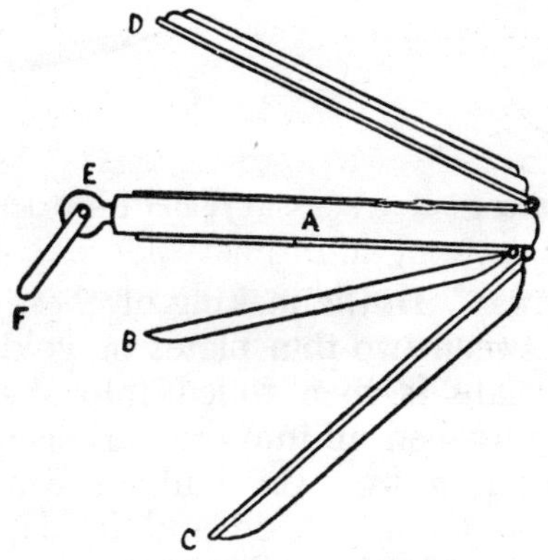

FIG. 166.—Parts of Watch Case.

watch clean if the wearer is engaged in a dusty business. The various parts of a case are: (Fig. 166) A, the "band," or middle, B, the "dome," C, the "bottom," D, the "bezel" that holds the glass, E, the "pendant," F, the "bow." A hunting or demi-hunting case also has a "cover" over the glass.

In many full-plate English watches the dome is a fixture and does not open. In other watches the dome is absent altogether; such cases are called "single bottom" cases, and are not to be recommended for wear. Bottoms and bezels are sometimes jointed to the case band and sometimes are loose, being merely snapped on tight. These circular snaps, as they are called, are much more dust tight than a joint can be. Some bottoms have fly-up springs and a lock spring, and are opened by pressing a "push piece" in the case pendant. These are convenient, but let the dirt in very much. After a little wear there is seldom any fit at all. Dirt also reaches the watch through the holes made for the push piece and the fly springs.

Many modern American watch cases have the bezel and bottom screwed on by a fine thread cut in a lathe. This is, perhaps, the best possible way of making a case. The edges of these are generally slightly milled to give a hold to the fingers in unscrewing them.

The Americans also have introduced the practice of making

Fig. 167.—Tightening a Bezel.

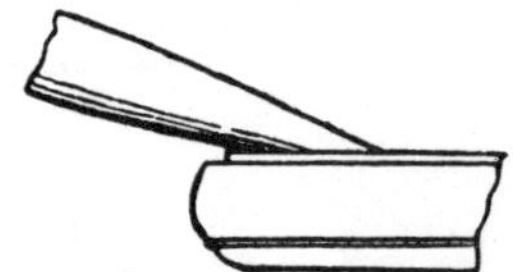

Fig. 168.—Tightening a Case Snap.

"gold filled" cases. In the making of these a plate of brass is sandwiched between two thin plates of gold and all brazed together. This plate is then rolled into sheet, pressed into shape, and finally worked up into the various parts of a watch case. The result is a case the bulk of which is brass, but which has all its surfaces, outside and inside, covered with a thin but sound layer of gold. These cases wear many years.

Snaps.—With wear the snaps of a case are apt to loosen. To tighten them, burnish the edge of the bezel or bottom inwards with a wetted oval burnisher and heavy pressure, as in Fig. 167. Or the snap on the case band, in the case of a bezel, can be burnished outwards, as in Fig. 168. A snap that is too tight can be eased by scraping the edge off the snap a little all round with a sharp square-edged steel scraper and afterwards burnishing smooth.

Glasses.—Bruises in a bezel on the edges of the glass groove can be pushed out from inside the groove with a watch screwdriver, and afterwards cut smooth inside by a graver point. A very badly bruised and damaged groove must be re-turned by a case-maker before it will hold a glass again.

Fig. 169 shows sections of the various kinds of glasses in use. A is a common "lunette," B a "double lunette," C a common "crystal," and D a "flat crystal." E is an old-fashioned "bull's eye" glass, and F a "patent hollow crystal."

Glasses are sized in a series of numbers subdivided into quarters and eighths. These numbers are based on the Paris "line" and the "millimetre." Lunette and double lunette glasses are generally sized in quarters; crystals and thin flat

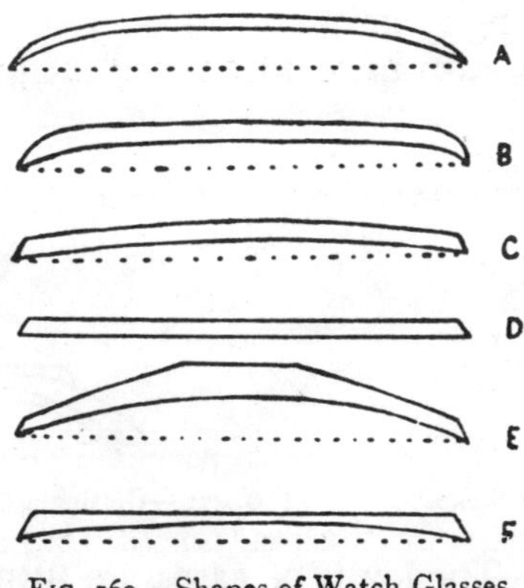

FIG. 169.—Shapes of Watch Glasses.

lunettes for hunters in eighths. Hunter glasses are also sized in "heights" as well; thus a $16\frac{5}{8}$ size may be had in 5, 6, 7, or 8 height. Each make of glass seems to differ a little, therefore it is best to stock one make only throughout, and fill up with the same to keep a good running series for fitting.

There is a knack in snapping in a good-fitting glass, only acquired by long practice. Never try to fit a glass with the bezel closed. A glass a trifle too loose may be fixed by running a drop of gum round the groove with a peg point, and turning the glass round to work it in; then leave an hour to dry. Do not attempt to tighten a glass by burnishing the groove, as the bezel will be ruined thereby. Similarly, do not try to cut a bezel groove out deeper to take a glass that will not quite go in. Instead, trim the edge of the glass with a No. 0 emery buff by hand, working gently round and round to keep it even; smooth off with a $\frac{3}{0}$ to finish.

Bruises.—A bruise in a dome or bottom may be knocked out by resting the inside on a boxwood case stake, like Fig. 170, and using a wood mallet. To prevent marking, cover the stake with tissue paper. Badly scratched cases are smoothed with rottenstone and oil on a leather buff or rag mop in a lathe, and polished with rouge and water on a mop, or by rouge on the hand. Bruises in the case band can be pushed or burnished out from inside and the remains smoothed off outside by filing, followed by smoothing with rottenstone and polishing with rouge.

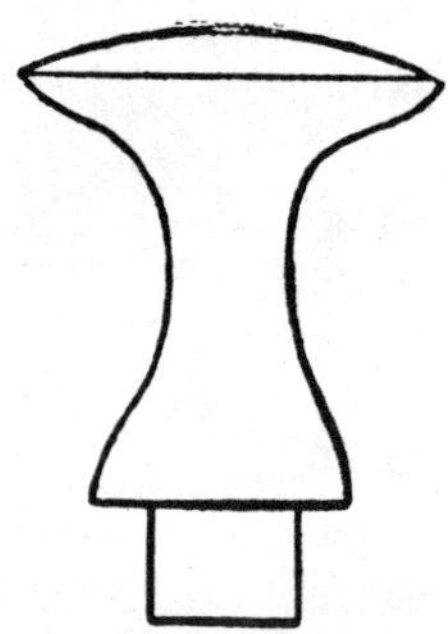

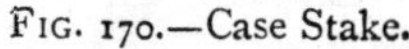

FIG. 170.—Case Stake.

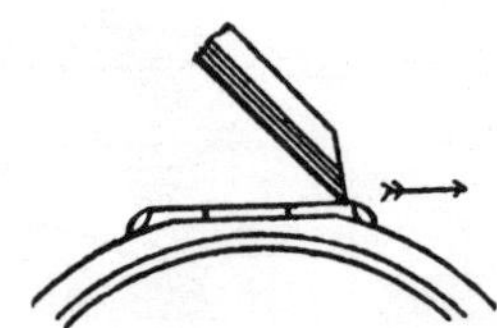

FIG. 171.—Getting a Plug out of a Case Joint.

Joints, etc.—The joints of cases are fitted with steel joint pins sunk in at each end. The end spaces are then filled in with silver or gold plugs and smoothed off neatly. To take off a jointed bezel or bottom, the silver or gold plugs can be picked or drawn out by inserting a graver point into them deeply and drawing them, as in Fig. 171. The steel joint pin can then be driven out. A joint that is "sprung," and will not allow the case to shut, can sometimes be sprung back by laying a thin wire along it inside and shutting.

Fly springs and lock springs are filed up from turned rings of steel, hardened and tempered. They are difficult to make, and, like new joints, are best left to case-makers.

A new push piece is a common repair. Push pieces are bought in the rough with a small piece of gold or silver soldered on the end. Hold by the thin end in a pin vice and reduce the head by filing until it passes into the pendant. Then reduce the thin end until it will go down and push the lock spring. Reduce it in length, and file out the hole through which the bow screw passes.

Bows.—Bow screws are always breaking or wearing out. They are bought by the gross and are easily fitted. Sometimes there is difficulty in getting the old one out. If anything at all projects, cutting nippers will generally turn the stump out. If nothing projects and it is tight, it must be drilled through. Pass the drill through the other hole in the bow as a guide to keep it straight. When drilled through (with a small drill), broach it out, and it will finally come out on the broach as a thin tube.

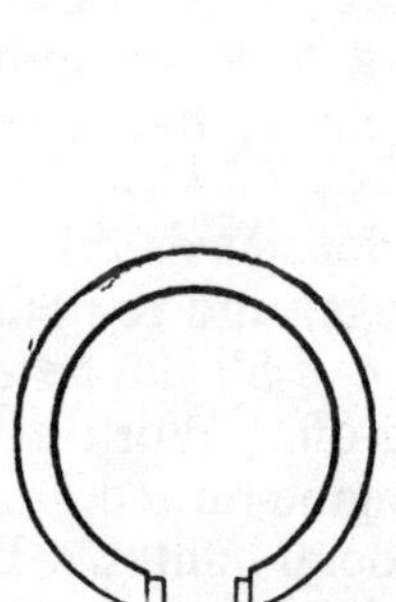

FIG. 172.—Keyless Bow.

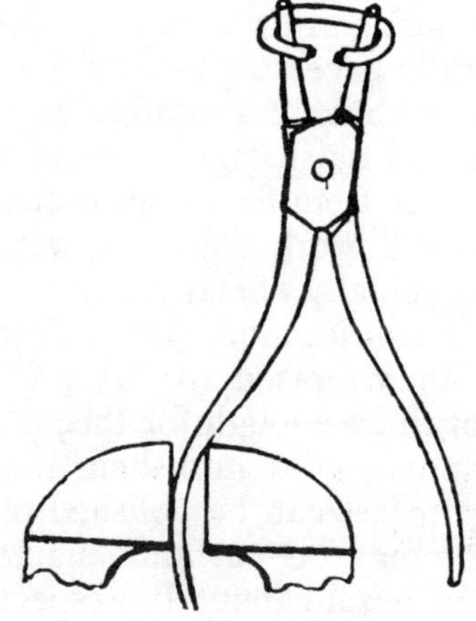

FIG. 173.—Opening a Keyless Bow.

The bow of a keyless watch requires pivoting in carefully, as in Fig. 172. A special tool is made for this purpose, but the operation can be done very passably by hand with a file, resting the bow on the filing block.

A case pendant in which the bow holes have worn very large requires plugging and soldering up by a case-maker, to make a good job, but a makeshift may be made by turning two washers or flanges to fit on the bow ends and lie in the worn hollows in the pendant.

The bow of a keyless watch can be sprung on or off by a pair of round-nosed pliers, as in Fig. 173, by screwing one side of the pliers in the vice and using the ends as a lever to open the bow.

Silver and gold are very soft metals, and burr when drilled. Therefore, in making a new keyhole in a case, use a very sharp drill and make only a small hole. Enlarge this by filing with a new sharp file, drawing it as required. Chamfer off the sharp edges with a rose cutter or a circular chamfering tool.

The dome and bottom often cause a watch to stop by pressing on the set-hand square, the fusee winding square, or the balance-cock jewelling. In these cases it is best to shorten the squares to free the case. But if it touches the balance jewelling the dome must be raised. Evidence of such contact is generally seen in the impression of the jewel screws or jewelling in the dome. Raise it by hammering with a mallet over a stake as in taking bruises out. When the dome has been raised it is generally necessary to raise the bottom also to free the dome.

The cover of a hunting watch is made to fly up by the push piece and fly springs. Sometimes the joint is stiff, preventing the cover from flying up properly. Run a little oil into the joint and work the cover vigorously for about five minutes. This generally effects a cure.

Cases with monograms, if not under 3 douzièmes thick, can have them erased by filing (save the filings) and re-polished. If not thick enough for this, a case-maker can fill the letters in by running gold into them and smoothing off. Worn engine-turned cases can be polished plain or re-engine-turned. A full hunter may be cut and enamelled as a demi-hunter. But a lunette bezel cannot be respectably made into a crystal. The case is always a bad shape afterwards.

CHAPTER XIV.

KEYLESS WORK.

KEYLESS watches, although very old, have only of late years become general. Comparatively few key-wind watches are made to-day, and fewer will be in the future.

The principle of all is that a milled "button" on the case pendant is connected with steel winding wheels, and winds the watch. Turning the button forward winds, turning it back does nothing, but it gathers ready for the next forward wind. The hand-setting arrangement differs very much. In most English watches, to set the hands, there is a "side push" to push in with the finger-nail. This throws the winding wheels out of gear and allows the hands to be set by the button. In "lever-set" watches, the bezel has to be opened and a lever pulled out to set the hands. In "pendant-set" watches, pulling the button smartly outwards throws it into gear with the hand-setting wheels. These are the principal methods, but variations of them are often found.

Rocking-bar Keyless Work.—Fig. 174 shows the "rocking-bar" keyless mechanism, as used in nearly all English, most American, and many Swiss watches. It is one of the simplest and most reliable.

The button is fixed on the outer end of a steel winding stem that runs in the case pendant. On the inner end of the stem there is a crown-shaped pinion, A, which drives the centre wheel of the rocking bar B. B drives both C and D. When driven forward C turns the ratchet E, which winds the mainspring up. When turned back to "gather," C slips past E, making a clicking noise. To set the hands, the side push F is pushed in. This moves the rocking bar bodily, the wheel C is lifted clear of the ratchet E, and D is thrown into gear with the minute wheel of the motion work, or with a small intermediate wheel, G, which runs on a stud free.

The rocking bar is sometimes made "lever set" by adding the lever H (Fig. 174). To set the hands, the outer end of H is pulled out and the inner end forces the rocking bar down, causing it to gear with the hand work.

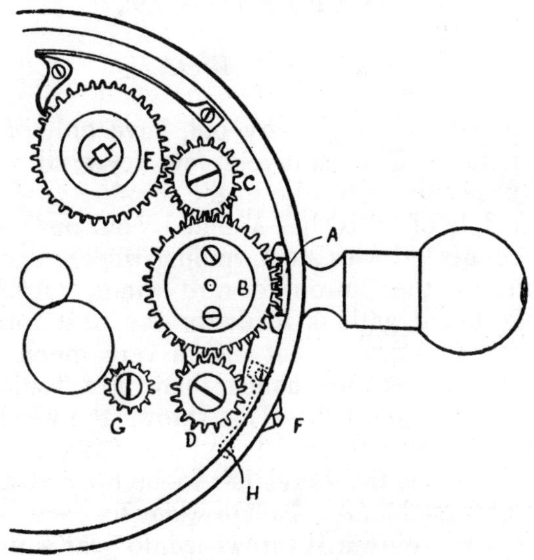

Fig. 174.—Rocking-bar Keyless Work.

In some watches the pinion A (Fig. 174) is fixed on a steel shaft running in bearings on the watch plate itself, and the button outside is fixed upon a key pipe which fits on the prolonged square arbor of the pinion A. In taking such a watch out of its case, a small set screw in the case pendant has to be removed, and the button and key pulled out; while in the pattern shown in Fig. 174 the button and winding pinion remain in the case, and the watch movement can be pushed straight out.

Rocking-bar keyless work has not many faults, but from wear and tear requires occasional repair. The button and winding stem sometimes run dry in the case, and stick, rust, or squeak. The button is sometimes screwed on to the stem, in which case, to remove it, hold the pinion A with pliers and unscrew the button with the fingers. More often the button is nutted on outside by a small gold nut sunk in its end. Hold the button in the fingers and turn the nut off with a screw-driver shaped like Fig. 175. Take out the stem and clean it; peg out the hole in the case, oil it well, and replace, making sure that the button is quite tight. It generally fits on its square best in one position, and that is marked by a file mark or dot on the square and a mark on the pipe of the button.

FIG. 175.—Pendant Nut Screwdriver.

A rocking bar that rises from the plate and allows the winding pinion depth to get too shallow can be tightened up by reducing the under side of the central steel boss of the wheel B (Fig. 174). This is of hard steel, and can only be reduced by laying it upon cork and using an emery stick, about No. 2. If the steel piece has steady pins this cannot be done, but the plate must be sunk a trifle by putting in the mandrel and taking a thin smooth facing cut across for the diameter of the steel only. The centre screw of a rocking bar, when there are no steady pins, has a left-hand thread.

If the teeth of the wheel C (Fig. 174) butt against the points of E, it shows that the click is not exactly in the right place. The click should be so placed that the wheel E presents *one* tooth to the wheel C. If two teeth are level C will butt. The remedy is to shorten the point of the click so that one tooth only faces C.

Broken teeth in rocking-bar wheels cannot be repaired, but must have new wheels. These are bought in the rough, and will require thinning by filing or turning, hardening, tempering blue, and the centres turning out to fit, in a step chuck, and recessing for the shoulder screws.

If there is an intermediate wheel between the rocking bar and the minute wheel, like G, Fig. 174, it must run quite free upon its stud. The stud is often a shoulder screw, and sometimes tightens the wheel up. If so, hold the screw by its tap in a split chuck and turn the head flange back a little with a thin graver point. This also applies to the wheels C and D on the rocking bar.

If a rocking bar is too tight, the centre steel piece of the wheel B may have a tissue-paper washer screwed between the plate and itself.

A brass minute wheel often gets its teeth broken through the pressure of setting the hands. They can be put in by slitting and soldering as before described, but are not very strong. The best remedy is a new *steel* minute wheel.

A rocking-bar watch sometimes winds badly because the depth between C and E is too deep. Like all depths, this one must have a *little* shake between the teeth. It is regulated variously in different watches; sometimes the bar, or a stud in it, banks on the plate, and the plate must be punched a little; or by a screw head, in which case a larger screw can be fitted.

Shifting-sleeve Keyless Work.—"Shifting-sleeve" keyless work is on a different principle altogether.

The diagrams (Fig. 176) show the upper and lower views

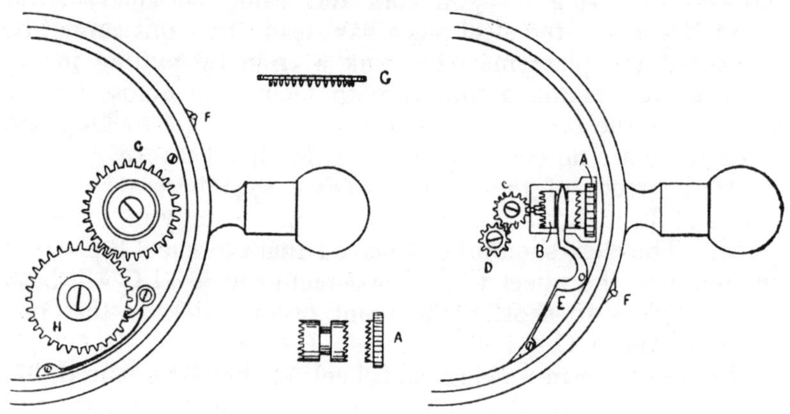

Fig. 176.—Shifting-sleeve Keyless Work.

of a typical Swiss watch with this kind of winding work. The button and winding stem pass right into the watch. The stem is squared, and carries a crown wheel or shifting sleeve, B. B can move upon the square stem by an easy sliding motion, but is bound to turn with it. A is the winding pinion, and rides loose upon the stem. It gears with the winding wheel G on the top plate, which in turn winds the spring by means of the ratchet wheel H. The pinion A has on its side or face next B a set of ratchet teeth engaging with a similar set on B. The spring E keeps B up against A. When the button is turned forward it carries B round. B carries A, and the spring is wound. When the button is turned backwards B slips past A, the spring E forming a click spring, and the result is only a gathering action. The side push F forces the spring E inwards, and carries the shifting sleeve B down to gear with C and D to set the hands, instead of driving A.

When well made, this is a very reliable form of winding work; but many of the watches to which it is applied are of the cheapest, and the workmanship is bad. The wheels A and B are liable to get the teeth worn or broken. They can be bought ready made in a finished condition. The spring E is liable to breakage. Rough ones can be sometimes bought, but they vary so much in shape and size that there is frequently no other way than to file up a new one from sheet steel and make it throughout. This job does not require any special instructions, as the old one or its parts are available for measurement. First drill the screw hole, and work from that as a fixed point. When roughed out, harden and temper it. Then finish down to size.

The depth between the pinion A and the wheel G is often faulty, generally shallow. When G is an ordinary wheel with one set of teeth, it can be deepened by getting A forward. This can be done by putting a very thin brass or steel washer behind A. When G has a double set of teeth, G must be got bodily down. If it merely has too much play, reduce the under side of its centre steel boss. But sometimes it must be sunk by turning the plate or bar in a mandrel. This keyless work wants plenty of oil on the winding stem and the ratchet teeth of A and B.

A new winding stem is easily made in a lathe. A piece of hardened and tempered rod is held in a split chuck, and the entire job done at one chucking, all except the squaring to fit B. Let it be finished smooth; polish the end pivot and the shoulder where A turns. The buttons are sometimes driven on to a squared end, sometimes screwed on, and occasionally held by a small side set screw.

The centre screw holding G has usually a left-hand thread.

Variations of this form of keyless work are found in American and a few English watches. In some the winding wheel G is not on the top plate, but under the dial, being screwed to a steel platform covering B and A. A lever-set arrangement may be fitted to it as to a rocking bar.

Pendant-set Arrangements.—The most important of these variations is, perhaps, the pendant-set arrangement adopted in many Swiss and American watches. Fig. 177 shows the arrangement adopted by Messrs. Patek Phillippe & Co., of Geneva, which is a typical one.

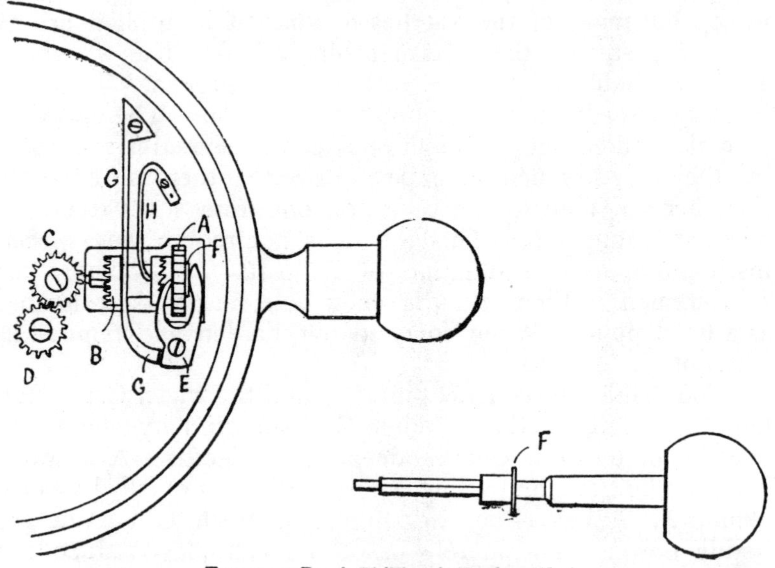

FIG. 177.—Patek Phillippe's Keyless Work.

The winding stem has a fixed flange, F, upon it, which comes just against A. A and B are of the same shape, as in Fig. 176. B is kept up to A by the spring H. G is another and stiffer spring, which in its normal position lies as in the figure. This watch winds exactly as described before; but to set the hands the button is pulled outwards. The flange F acts on the steel piece E, which draws the pinion A, and keeps it against F. E also forces G downwards, and G carries B down with it into gear with C and D, the motion wheels.

These watches of Patek Phillippe unfortunately require taking partially to pieces to get them out of their cases; but the keyless work is well made and very reliable. By varying the arrangement a little, as in "Omega" lever watches and some others, including some patterns of Walthams, the flange F, Fig. 177, is dispensed with, and a separate piece, A, Figs.

178 and 179, added, operated by a *slot* or *groove* in the winding stem. This enables the watch to be got out of its case more easily, as by loosening the screw holding the separate piece it is liberated from the groove in the winding stem, and the latter can be drawn out. Figs. 178 and 179 show two forms of this kind.

The separate parts of this keyless work, winding stem, springs, etc., have to be filed up and made by the workman if

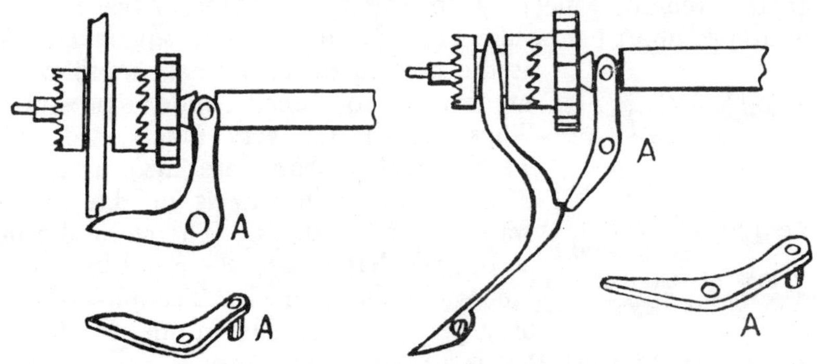

FIGS. 178 and 179.—Forms of Pendant-set Keyless Work.

they want replacing. But the wheels can be bought finished in most cases.

Some American pendant-set watches are very complicated, and have many curiously shaped springs difficult to make. Fortunately, they can be purchased from the English agents of the manufacturers. The idea of American manufacturers is to produce a watch which shall be interchangeable with any standard case. The form of pendant-set keyless work above described does not lend itself to this very well. Hence their efforts have been directed to devising a mechanism that shall be all contained in the watch, and not depend on the winding stem or case. The idea has generally been carried out by making the winding stem hollow like a tube, and letting it lie in bearings in the watch plate. Pulling the button outwards allows a central shaft in the hollow stem to slide up, and by an arrangement of levers and springs, slides B into gear with the motion work. In their later watches these systems have been abandoned for simpler mechanism more resembling Fig. 177.

In these and many other keyless watches slender steel

springs like fine wire, hardened and tempered, are often used. They frequently break, and sometimes are lost. Often they are troublesome to make, and in many cases springs answering the purpose can be quickly made from fine brass wire, hammered stiff and bent up to shape. Such brass springs generally act as well as the original steel ones, and are less liable to break.

Winding Wheels.—Keyless winding wheels require great strength, especially in the teeth. Most have round-bottomed short teeth, like Fig. 180 at A. The square angles at the bottom of an ordinary tooth offer invitations to break under pressure. These teeth are very much stronger. Teeth of this shape are also used for barrels and main wheels in the going train in good watches. The teeth shown at B, cam-shaped, are also used in many Geneva winding wheels. Teeth like these only permit of running in one direction, and in keyless work that is all that is necessary.

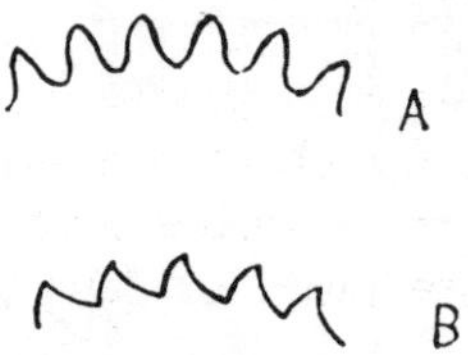

A

B

Fig. 180.—Shape of Winding Teeth.

The forms of keyless work shown here are the usual ones met with; but there are a few others occasionally seen, especially in hand-made English watches. Although differing in arrangement, they will present no special difficulty to the repairer.

Fusee Keyless Work.—All these different kinds of keyless work are applicable only to going-barrel watches. Fusee watches, if required to be keyless, present a difficulty.

In the first-mentioned class—the going-barrel watch—the barrel arbor stands still while the watch goes, and only moves when turned by the keyless work in the act of winding. This simplifies matters, as the winding work can always be in gear with the barrel arbor. In the fusee watch the fusee square is always slowly turning round while the watch is going, and therefore the winding work and button must be free and disconnected from the fusee, except while winding. If not, the whole keyless work would revolve while the watch was going, and the button would catch in the pocket and stop the watch.

In the rocking-bar watch it has been seen that the ratchet must present *one* tooth to the winding wheel, or butting of the teeth will result. But a wheel on the fusee arbor is liable to be in any position at the moment of winding, and therefore nearly all fusee keyless watches butt badly as they are wound. In some there is a kind of rocking bar, which normally is disconnected with the winding and hand work. To wind, a push is held in, throwing the rocking bar into gear with the fusee wheel. To set hands, another push is needed. Or one push may be made to answer both purposes by being pushed in to set hands, and let out (by opening the case back) to wind. Others wind to the left, and throw into gear automatically with a sliding motion of the rocking bar, and set hands with a side push. A few wind with a circular gathering click working round a ratchet, throwing itself into gear when pressed forward. These automatic arrangements all have an uncertain loose feel in winding, and waste much power; while the positive action ones (by opening case, side push, etc.) are more or less a nuisance, and butt badly as well.

Altogether, fusee keyless watches are unsatisfactory and troublesome, not one in twenty winding even passably.

CHAPTER XV.

CAUSES OF STOPPAGE OF WATCHES.

Small Faults.—A watch may be cleaned carefully and well repaired, yet may, when put together, have a bad action, or stop in the workman's hands or under his very eyes. Such watches often give more trouble and waste more time than has been previously spent on cleaning and repairing them.

Beginners, after they have vainly looked all round such a watch, are apt to give it up as a "mystery." There are no mysteries in watchwork. A stopping watch has a fault somewhere, however small, and it must be found.

A workman who looks out for faults as he takes a watch apart is not troubled with these "stoppers" so often as the careless man. An escapement should always be examined and tried before taking apart; all wheels should be examined as to endshake and side play, marks of fouling each other or the plates looked for. As the watch is put together again, everything should be tested as it is put in.

Often the cause of stoppage is a dial pin fouling a train wheel, or a screw put in its wrong place, its point going too far through somewhere. Stoppages in the train are most easily found. Taking care not to start such a watch, with a needle try the scape wheel and see if it has "power" on it. If not, try the fourth, third, centre, and so on in succession, until the point is found at which the power disappears. This gives a clue, and a bent tooth, a bent pivot, a tight endshake, or some cause of fouling must be looked for in that wheel or pinion. A watch that stops once per minute probably has a fault in the fourth wheel or pinion. If once per hour, it may be the centre wheel or pinion or the motion work. If every fifteen beats, a damaged scape tooth or dirt in the scape pinion.

When power appears to go off at the barrel itself, it may be that the stopwork jams. A little roughness on the centre stop finger of Geneva stopwork is liable to catch the points of the star wheel. A shallow depth between stop finger and star wheel will also cause it to jam. The mainspring may bind in the barrel, or a cap screw point may be too long and bind it.

If a fault is suspected in the train and nothing can be seen to account for it, run each wheel separately in its frame to see that it has perfect freedom; then each two wheels together, and try all depths.

A watch that seems to have plenty of power but has no action has a fault in the escapement. After trying the scape depth, the banking shakes, run, and endshakes, see that the lever and roller do not touch, that there is no oil between the lever and the plate, that the hairspring lies flat and true, free of the balance, the plate, and everything else. If no fault can be found there, run the balance in its pivot holes alone, with the roller and hairspring removed. Let it run slowly, and observe how it stops. A bristle from the brush may be found sticking in its rim, or in the plate, just touching it. The balance rim may touch the fusee chain in a $\frac{3}{4}$ plate. Then put on the roller and let the watch go to half time, without the hairspring, and see if it acts freely and does not catch anywhere.

A pivot may not come quite through a jewel hole, or the hairspring collet may touch the balance cock.

Much can be learned by listening carefully to the beat in various positions. If the balance is foul of anything, a striking will be heard. If a Geneva horizontal scrapes in one position, it is probable that the scape wheel touches either the bottom or top of its passage in the cylinder.

A watch that stops in one position only has generally a fault in the escapement. If the pivots and jewel holes are quite right, it is a fault caused by the movement of the balance, scape wheel, or pallets, owing to their endshake or side play. Something that is quite free with the endshakes one way fouls when they are the other way. Excessive endshakes may be present.

Note if the action is equal dial up and dial down. Any falling off will indicate a fault. To observe the action dial up, a small piece of mirror laid upon the bench is extremely useful, as the watch can be held steadily over it and the action observed at leisure.

A lever watch that falls off in action when the balance leans towards the lever has probably a fault in the roller and lever depth. The ruby pin may be a round one and not enter the notch properly. A flatted pin will cure this. There may be not enough, or too much, banking shake, or a roughness on the roller edge.

Watches that go all right lying down and hanging up, but stop in the pocket, will often be found to be too shallow in the safety-pin action. The safety pin in the lever can jam against the roller edge. Also, in such a case, look to the pallet depth and see if it mis-locks on any teeth, as this causes the safety action to jam.

If a watch stops when just wound up, see to the maintaining work, also that it is perfectly in beat and that the safety action is correct.

In full-plate English watches the barrel may rise just a little above the plate and foul the under corner of the balance cock, or the balance rim, in certain positions. If the mainspring is of the American pattern and has a brace, the top pivot of the brace may project and foul the balance rim or scrape the barrel bar. The point of the guard pin in the lever may touch the under side of an undersprung index.

The chains of these watches may scrape the back of the potance, or the inside of the cap. The barrels of $\frac{3}{4}$-plate going-barrel watches may just touch the case edge and cause binding. The cannon pinion may be too low, and its teeth may scrape the plate.

The dial may press upon and bind the minute wheel or some of the points of the lower pivots.

Hands should always be looked at to see if they touch the dial or the glass and are free in the dial holes.

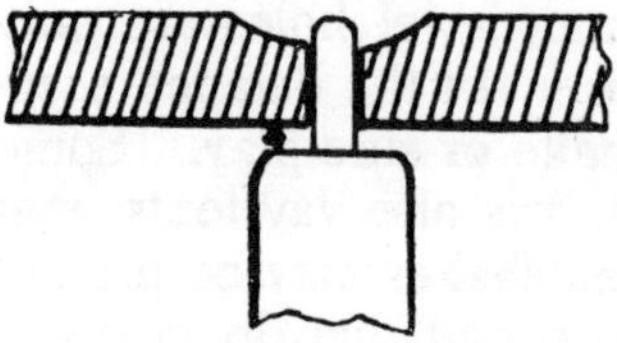

FIG. 181.—Grit jammed between Pivot Shoulder and Plate.

Some new machine-made English levers give a lot of trouble by stopping from the least dirt. It gets round the train-wheel pivots and jams them tight. This is caused by rough plates and rough pivots. Polish all the pivots and stone the plates round the pivot holes quite smooth. The gilding on the plates of these watches is like a honeycomb.

The plates seem to be pickled to frost them deeply before gilding, presumably to cover up a want of smooth finish. The result is that grit is held by the surfaces of the plates and jams under the square pivot shoulders of the train wheels, as in Fig. 181.

If dial pins are not pushed in quite tight and do not fit well, they often work loose and come out. Being in the case, they rattle about until they get under one of the wheels and there stick. When a stopping watch is brought in minus one dial foot pin, look at once to see if it is not somewhere in the wheelwork. Similarly a small screw, if it works out, may jam the wheels.

A train wheel or pallet pivot that is short and does not come through its brass hole is apt, in an old watch, to wear a step in the hole. When much worn in this way, the step sometimes jams the pivot by stopping its endshake, as in Fig. 182. The remedy is to broach out and bush the hole.

In the holiday season, when people go to the seaside, *sand* is a frequent cause of watches stopping. A little gets in the pocket and finds its way into the watch case. One grain is sufficient to stop it, though, as a rule, many grains find their way in. A single grain fixed between two wheel teeth as a rule causes the stoppage.

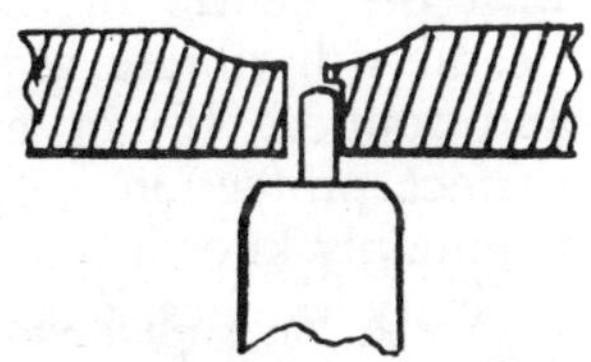

FIG. 182.—Worn Pivot Hole.

Duplex watches and pocket chronometers, even when set in beat as carefully as possible, will occasionally stop in the pocket. Being single-beat escapements, it only needs a movement that brings the balance to rest at the right moment to stop them.

Verge conversions that have no maintaining work will also stop during winding sometimes. These should always be very carefully set in beat, to avoid this trouble as far as possible.

Watches with heavy escapements are also liable to stop when the hands are put back, especially if they move stiffly. For this reason, pocket chronometers should always have the hand work left easy, just tight enough for them to carry with certainty, but no tighter.

In some old English $\frac{3}{4}$ plates the lever lies much too close to the pallet cock—so close that the oil applied to the top pallet pivot is almost sure to be drawn between the lever and the cock. The best way to serve them is to file away the under side of the cock each side of the pivot hole, and, if possible, turn a cone hollow around the top pivot in the body of the pallets.

Magnetized Watches.—In these times, when there is so much electrical machinery about, watches are often found to be magnetized.

If all the steel parts should be very strongly magnetic, the watch will not go at all, the balance being attracted by the screws, etc., around it, and the hairspring adhering to the balance. When only slightly magnetized, the watch will go, but the timekeeping will be very erratic. A watch becomes magnetized through being brought into the field of a powerful electro-magnet, such as a running dynamo or motor. To detect it, place a small charm or pocket compass flat over the balance cock, with the centre of the compass corresponding with the centre of the balance. If magnetic, the compass needle will vibrate as the balance does, or perhaps fly round and round. If not magnetic, the needle will be quiet. A perfect protection to a watch is an iron box, or what is commonly known as a "tin" box. It follows from this that ordinary watches, if in "gun-metal" cases (oxidized or blacked iron), and full hunters are perfectly protected from magnetic influence.

To demagnetize a watch, it should be revolved rapidly and brought into the field of a powerful magnet—say a dynamo—and gradually withdrawn. One way of effecting this is to first wedge the balance with tissue paper; then fasten a string to the bow and twist it up tight; suspend the watch by the twisted string close to a running dynamo, and let it untwist rapidly. As it does so, withdraw it out of reach. This is only a rough way. If a magnetized watch is sent to an electrician, he will demagnetize it more thoroughly for a fee of about 2*s.* 6*d.*

Careful tests show that once a watch is magnetized it is never *quite* free again. A fine watch with a close rate is for ever spoiled. The only perfect cure is to heat all the steel parts to redness, re-harden, and polish them again. This involves fitting a new balance, hairspring, mainspring, etc., and in most watches is not worth doing.

Karrusel Watches.—These watches have a few faults of their own that often stop them. The revolving carriage carrying the escapement must be dry, clean, and quite free. Where it bears and rubs on the plate, both top and bottom, must be channelled out by turning, letting it bear on circles only. These must also have no gilding on them, being stoned off smooth where the friction comes. Gilding wears off and works up into a black powder and causes choking.

The edge of the carriage is sometimes found to foul the end of the third wheel cock and the tops of the barrel teeth. The third cock is easily filed to free it. To free the barrel teeth, the barrel is best put in a step chuck and the teeth-tops bevelled down. The top third pivot hole is also rather liable to be left dry, and must be oiled before putting in the third wheel.

Bad Mainsprings.—A watch that has a poor action, the balance vibrating less than one complete turn, is always liable to stop from the least cause. Often such a watch will be found to have a poor, cramped up mainspring, and can be greatly improved by changing it for a good quality *lively* spring of the same thickness. Nothing pays better in watch materials than to keep a good quality of mainspring. A spring costing 4*s*. 6*d*. or 6*s*. per dozen, instead of 2*s*. 6*d*., will pay for itself many times over in making poor watches go well.

Unequal Rates.—Some watches after cleaning are found to go much slower hanging up than when lying. Generally this indicates a balance out of poise, or a hairspring not true in the centre. In very small ladies' horizontal watches this fault is often troublesome, and may be sometimes cured by deepening the cylinder depth, *i.e.* setting the cylinder closer to the 'scape wheel. In these very small and cheap watches side play of pivots causes great inequality of action.

CHAPTER XVI.

CONVERSIONS AND ALTERATIONS IN MOVEMENTS.

A VERGE with a good frame, good train wheels, and a sound case often pays to convert to a lever. An English cylinder or duplex, when the escapement has become worn, may be so altered. A pocket chronometer that has become badly damaged in the escapement will often cost many pounds to restore, and still be no better and not so strong as if converted to a lever at a quarter of the cost.

Making the New Escapement.—These are all cases in which conversion is advisable. The task is not a formidable one, and can be undertaken by any fairly good workman. An English full-plate cylinder, duplex or chronometer, is the most easy to alter. In these cases the pallet staff may be fitted to the old scape-wheel holes and the jewelling may remain as

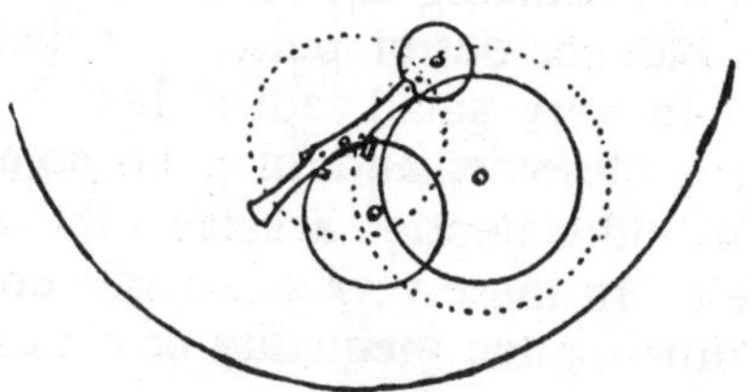

FIG. 183.—Full-plate Duplex Conversion.

before. A new fourth wheel of a smaller diameter may be fitted, and the new scape wheel and pinion pitched as near to a right angle as possible, like Fig. 183, in which the dotted circles show the position of the old fourth wheel and scape wheel.

Scape wheels and pallets are bought in pairs matched to

each other. Levers and rollers similarly matched can also be bought, but in this case, as the pallet holes are fixed, a lever and roller must be specially made to the depth. The distance from the pallet holes to the balance holes must be measured and divided into four parts. One-fourth should be the distance from the balance staff to the ruby pin hole in the roller; the other three parts should equal from the pallet hole in the lever to the notch. This gives a ratio of lever to roller of 3 to 1.

First, procure a rough roller, or make one from steel rod. Measure and mark the ruby pin hole. Drill it and broach it true. Ordinary flatted pins are fitted in round holes. A D-shaped, half-round pin, an oval, or a triangular pin, require holes in the roller to fit them. A round hole is first drilled; then a hard steel punch of the right shape is driven in the hole. The roller edge in front of the hole can be hammered in a little to ease the punch and flatten the front of the hole. The roller can then be hardened and tempered, and turned true and flat. It should be reduced in diameter until there is a little more steel outside the pin hole than is necessary for the passing hollow. The surplus is to be trimmed off at a later stage. The hollow can be filed.

To make the lever, take a strip of lever steel and cut off a short length. Drill the centre hole, and measure and mark the position of the notch. Drill a hole there and file into it, slanting on each side to form the "horns," as in Fig. 184. With a slitting file cut the notch, and open its sides true and square with a notch "side file."

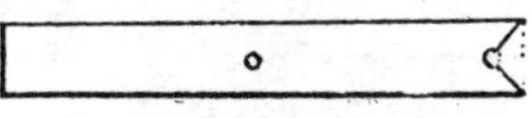

FIG. 184.—Making a Lever.

Fit a brass pin in the roller and put roller and lever in the depth tool on turning arbors. Set the tool to the depth from the pivot holes in the plate. Open out the notch to depth and width. Mark and drill the guard pin hole as small as possible, and close to the bottom of the notch.

The pallet staff can be turned and fitted to the pallets and pivoted, the balance staff made, the balance mounted, the scape pinion turned and pivoted, and the wheel mounted, colleting it with gold, as this metal will enable a good rivet to be burnished on it. Broach out the centre hole of the lever to go tightly on the pallet staff up to the pallets. It is important that it should be quite tight. Put the pallets and scape wheel in the depth tool and adjust to depth accurately, leaving it *just the least shade too deep*. Strike this circle from the pallet hole as a centre. Put the scape pinion and fourth wheel in the depth tool, and strike another circle from the fourth wheel hole as a centre. The intersection of these circles marks the point for the top scape pivot. Drill the pivot hole, upright it in the mandrel, and mark and drill the bottom hole. Run in the scape wheel. Then put in the lever and pallets, scape wheel, and balance. Turn the lever round upon the pallet staff until it is "in angle." That is, the teeth must "drop" on the pallets at equal distances on each side of the line of centres. If the roller edge is trimmed down by trial, until when a tooth drops the guard pin just touches its edge, there will be no difficulty in getting the escapement in angle. When correct, drill the lever and pallets and pin them together. This done, the lever may be shaped up, hardened and tempered, its flats and edges polished, and the notch opened out to fit the ruby pin and polished inside. The roller can be polished on flats and edges. The flat can be done overhand on a brass polishing block. The edges may be polished in the lathe or turns. An escapement maker would use swing tools for this polishing, but the average repairer does not possess these, and would not have sufficient practice to use them properly.

The scape depth must be carefully tried to see how deep it locks, and the wheel topped in the turns until it locks as lightly as possible.

This depth is a very important one, and a minute error in drilling the pivot holes or in the registering of the depth tool is fatal to it. This is why it is advisable to pitch it just a shade deep, as the wheel can be topped to shallow it as required. Most depth tools have an error; that is, the outside points do not accurately register the same as the inside centres. A good plan with a faulty depth tool is to reverse the runners with the points *inside*, and score the plate with the points in that position.

To mark the points for the banking pins, put the balance-staff and roller in, lay the lever on the top plate with the pallet-staff pivot in the top hole, and with the guard pin on either side of the roller mark just at the lever edge. Drill them a little closer than marked, and trim down the lever edges until the bankings are correct. Do not forget to poise the lever.

Three-quarter-plate Watches.—A $\frac{3}{4}$-plate duplex or chronometer generally requires a new scape cock, as the old one is seldom the right size and shape to take the wheel and pallets. In these the new scape pinion can be run in the same holes as before, or the old pinion can sometimes be used again. If a very small wheel and pallets are used, a right-angled escapement can be made, though sometimes this is not possible. In such a case the lever has to be "dog leg," like Fig. 185, to accommodate the pins. For $\frac{3}{4}$ plates a lever and roller to match may be bought, and the position of the pallet holes arranged to suit them; or they may be made as before described. The arrangement of such a conversion is like Fig. 186, the dotted circle showing the old scape wheel.

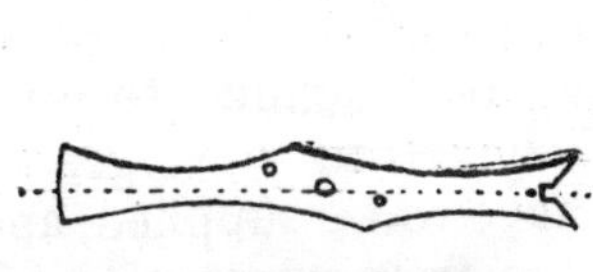

Fig. 185.—Dog-leg Lever.

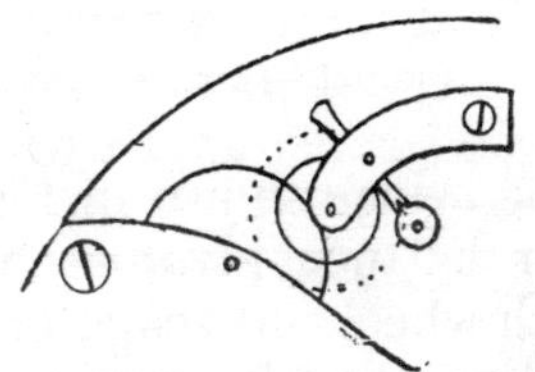

Fig. 186.—$\frac{3}{4}$-plate Duplex Conversion.

In converting chronometers, duplex and cylinder watches, new fourth wheels, where required, and new scape pinions may be of the same numbers as before. Then the hairsprings will be used again, together with the balances, just as they are. But if such a watch has a plain balance, it will be much the best to put on a new compensation balance, if there is room for it. In this case it will need re-springing as well.

Converting Verges.—Verge conversions are rather more trouble. Some verge trains are so arranged that the third

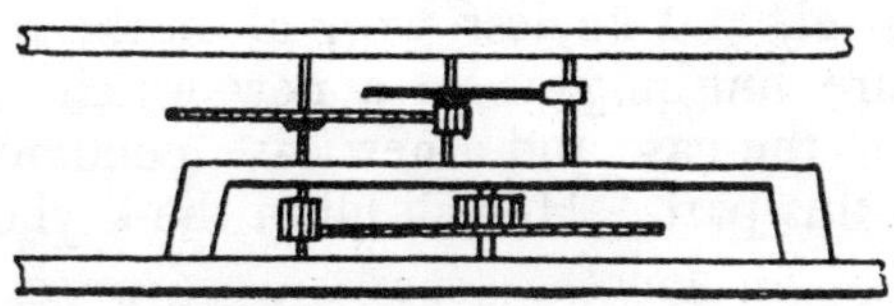

Fig. 187.—Arrangement of Verge Conversion by a Bridge.

pinion can be left alone and the escapement planted quite free of it. But the majority require a platform made to go over the third wheel, like a bridge, and carry the escapement. The third pinion then must be cut down very short and its top pivot run in the bridge. The fourth pinion comes through a hole in it. The arrangement is shown in Fig. 187. The old potance can be generally filed and turned out to take the roller and allow the lever to reach it. If not, and a new potance has to be made, the entire escapement may as well be included in it, and the top third pivot run in its under side, as in Fig. 188;

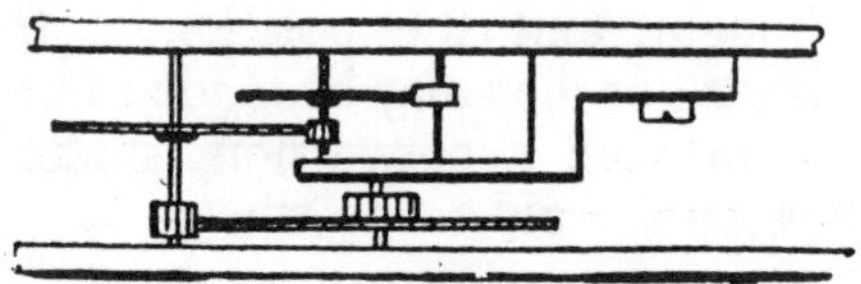

FIG. 188.—Arrangement of Verge Conversion by a Large Potance.

or the bridge shown in Fig. 187 may be extended to take the lower balance pivot as well. In converting a verge, a new fourth wheel and scape pinion will have to be supplied, and the numbers must be such as to make the train as near to 16,200 as possible. They can be calculated according to the directions given on p. 133.

An old "club roller" or "rack lever" is easily converted by supplying a new wheel and pallets, and lever and roller, and re-pitching the depth.

Keyless Conversions.—A really good $\frac{3}{4}$-plate key-wind fusee English lever is sometimes worth converting into a keyless. The best job is made by dispensing with the fusee and fitting a going barrel, with a new top plate, fitting ordinary rocking-bar keyless work and new motion work. The movement, if sent to a movement maker, will be fitted with top plate, new barrel, and keyless work all in the rough. It will then all require finishing as in a new watch. It will want "boxing in" to the case and a new case pendant. The casemaker will do this part. He will fill in the keyholes also.

Improving Watches.—A good watch with compensation balance and a flat spring may often with advantage be altered to a breguet. Occasionally the same balance cock will do again, but generally it will be best to make a new one and stud it on the left, fitting a proper index. The old spring, if a good one, may be turned up into a breguet, or a new best hardened and tempered one put on.

Many a good $\frac{3}{4}$-plate English lever may be much improved and made to look many years newer by a small outlay. Top plates of the shape shown in Fig. 189 may be filed up to a nice curve, as shown by the dotted line, and a balance cock like Fig. 190, A, may be turned like B. Plates may be stoned

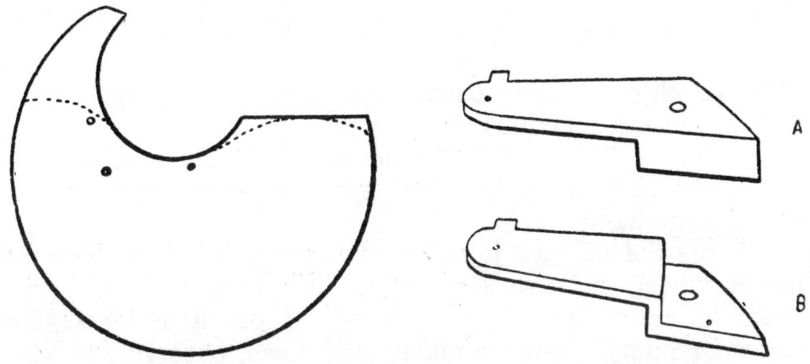

FIG. 189.—Altering a Top Plate. FIG. 190.—Altering a Balance Cock.

smooth and free from scratches, and the frame re-gilt. The squares, index, etc., can be re-polished and the screws re-blued. Scratchy blue screws may be re-blued without polishing, as may hands and other blue steel parts that have partially worn bright. Brush them clean and free from finger marks, and place on the blueing slip until the blue reappears all over.

The tops of brass jewel settings get shabby, or after stoning and gilding a plate, stand up too high. In such cases stone them smooth and level as they lie in the plate, then remove and polish them upon a pewter polishing block that has been filed flat. Apply red-stuff and oil and insert a peg point in the jewel hole to rub them over the block with circular strokes.

CHAPTER XVII.

MARINE CHRONOMETERS.

MARINE chronometers require very careful handling. Take the movement, together with the inner brass box, out of its gymballing. To take the movement out of the box, unscrew the bezel and turn the box upside down, receiving the movement in the hand.

Before the balance is removed insert an oiler to block the fourth wheel, in case the detent is accidentally touched. Remove the balance from the cock and put it under a glass cover for safety. Let the mainspring down, observing to what point it was "set up." The ratchet is generally marked to go on the right square, and the teeth are marked where the click point comes. Free the maintaining detent from the ratchet so that all power is off. Then take out the detent and its banking arm. Remove the hands, dial, motion work, and the small pinion on the fusee bottom pivot (the latter with brass-nosed pliers; it is pushed on only). Take off the cup around the fusee winding square, and take out the barrel and bar. The plates may then be taken apart. Great care must be taken of the scape wheel. It may be put under another glass cover with the detent.

All pivots should be examined for roughness or signs of wear. Cut or rough pivots should be re-polished, and every pivot must be perfect. When re-polished, those running in brass holes should have the holes bushed. Those that run in jewels must have new holes fitted by a chronometer jeweller. If a scape or balance pivot is so cut that it will be too thin when polished, shorten it, re-turn and polish, and send the frame and arbor to the jeweller to have the jewelling sunk further in the plate or cock to suit the new pivot.

Polish the frame and cocks with rouge and oil on the finger-tips by rubbing all over, not forgetting the edges. Clean off with benzine on a watch brush, and brush clean and dry as a watch. Take the fusee apart, clean and oil its internal parts. Wipe out the mainspring and re-oil with best French clock oil.

In putting together, first screw on the brass edge to the pillar plate. It can then be well handled without fingering the polished parts. Oil the fusee and bottom centre pivots with the clock oil, the rest with chronometer oil. Brush the balance clean, clean its ruby pallets with a dry peg point, and put in place. Clean the detent by laying it flat on tissue paper and applying a peg point to remove dirt from the locking jewel face (see that this is clean and bright) and the gold spring point. Put the detent and its banking arm in last of all, after the chain has been put on, the spring set up, and the balance put in. Put it in its brass box, and proceed to put on the motion work, dial, and hands. Wind it up, seeing that the chain runs true upon the fusee, and put on the "up and down" hand at "up."

See that oil is put to the maintaining detent pivots, the point where the spring touches it, the stop joint, and the stop spring. Put no oil on the pallets, the scape-wheel teeth, or the detent.

Put in its gymballing again and oil the pivots; see that when hanging free it is quite level. There are adjusting screws to level it in both directions.

In putting on the seconds hand, see that it falls truly on to the seconds mark, and stop the chronometer before putting it on, or damage to the scape-wheel teeth may result.

Rust on any steel work must be scraped out with a sharp graver. On the balance spring it is fatal, and a new spring must be applied.

Adjusting and Rating.—For a chronometer to have a good steady rate, it must be compensated truly and the spring must be isochronous. The long and short arcs may be tried by letting the mainspring down a turn and setting it up again, testing several days in succession. The balance must be in perfect poise, for although the "position rates" are absent, a want of poise throws a side strain on the pivots and introduces error in the rate.

For loosening the screws that hold the balance weights, the

spring must be removed and a screwdriver with a blade like Fig. 191 used. This should be arched enough to turn well over the top balance staff pivot, and long enough in the arch to span over the opposite weight. It is always well to make a slight mark on the rim before moving a weight, as once the screw is loosened the position is apt to be lost sight of.

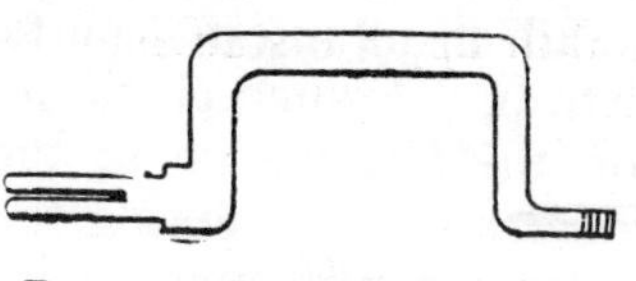

FIG. 191.—Chronometer Balance Screwdriver.

If a chronometer is carried or sent upon a journey, it is well to wedge the balance firm with two thin cork wedges inserted under the rim, one on either side.

Adjusting the Mainspring.—To ensure that the second day's rate of a two-day chronometer is the same as the first, or that the successive days of an eight-day chronometer are equal, an adjusting rod must be used to test the force of the spring when wound up and when run down.

A simple form of adjusting rod is illustrated in Fig. 192.

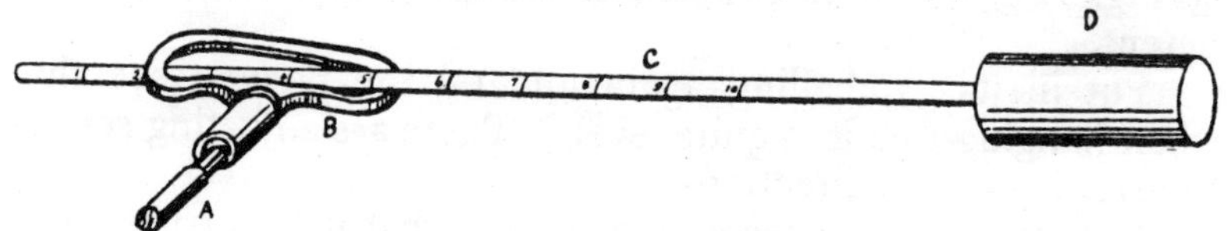

FIG. 192.—Simple Form of Adjusting Rod.

A = fusee square, B = key, C = adjusting rod, D = balance weight. It consists of a steel rod 15 in. long with a fixed weight of $\frac{1}{2}$ lb. on one end. It is put through a key handle as shown, and shifted until it balances the pull of the spring. If tried thus, any difference in pull between "up" and "down" can be noted by a scale marked on the rod. If too strong when wound up, the spring must be set up a little more. If strongest when nearly run down, the spring must be set up less.

New Pivots.—A new pivot is best put in a balance staff or a scape pinion, as shown in Fig. 193. The staff or pinion is cut off where marked at C, and drilled as at B, with as large a hole as possible. A plug is turned in a split chuck from tempered steel, as at A, and driven home. A pivot put in thus will be as good as new and defy detection. The cone shoulder can be made as large as before. To drill up, the staff may be

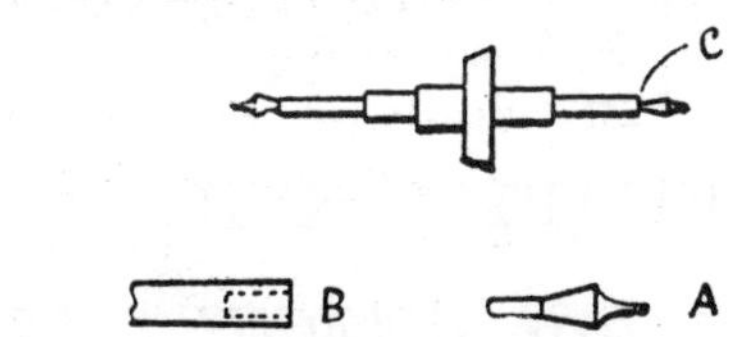

FIG. 193.—New Pivot in Marine Chronometer Balance Staff.

held by one end in a split chuck, and the other may be steadied by coming through a hole in the back cone plate runner.

Chronometers that have run a year or two generally have a hard sediment caked on the face of the locking stone and on the impulse pallet. This is probably dirt from the scape teeth or fine particles of brass driven upon the polished stones by constant little blows, until they stick hard. When cleaning, these jewel surfaces must be carefully scraped clean, if necessary, with a graver point, and left perfectly smooth.

CHAPTER XVIII.

REPEATING WATCHES, ETC.

REPEATING watches strike the hour, quarter, etc., last indicated by the hands, upon pressing the pendant or pulling a slide. Small steel hammers strike circular wire gongs arranged round the movement just inside the case. There are generally two gongs fixed on one base, or it may be on separate bases. Fig. 194 shows a Swiss repeater movement in which the hammers and gongs can be seen in position.

FIG. 194.—Swiss Repeater Movement.

Quarter repeaters strike the hour on one gong and the quarters on two gongs, known as ting-tang quarters. *Half-quarter repeaters,* in addition, strike one on the second gong after the quarters, if the half quarter has passed. *Minute repeaters* strike the hour on the first gong, ting-tang quarters on both, and the minutes on the second gong. *Five-minute repeaters* strike the hour on the first gong, and the number of five minutes past the hour, up to eleven, on the second gong.

Clock watches strike the hours and quarters like a quarter-chiming clock as they come round, constantly reminding the wearer of the time. In addition, they will repeat like a minute repeater when required.

Repeating Mechanism.—Repeating watches contain a second small mainspring to do the striking. The act of pushing in the pendant or pulling the slide winds the spring up a little. In running down it operates the hammers, through racks with saw-shaped teeth, and is regulated by a small train of running wheels and a fly pinion, or an anchor escapement like an alarm clock, to prevent it running too fast.

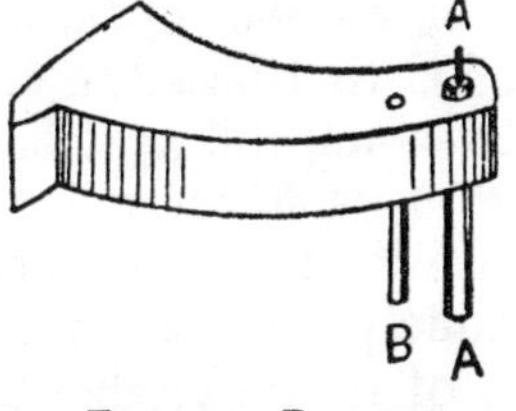

FIG. 195.—Repeater Hammer.

Hour-striking Work.—Fig. 195 shows a repeater hammer. A, A are its pivots upon which it turns; B is a steel stud that passes through a slot in the plate and by which the hammer is operated.

Fig. 196 shows the under side of the watch plate. The pivot A comes through a pivot hole and stands up; B comes through a slot; C is the hour rack mounted on the squared arbor of the main wheel of the repeating train; D is the hour hammer

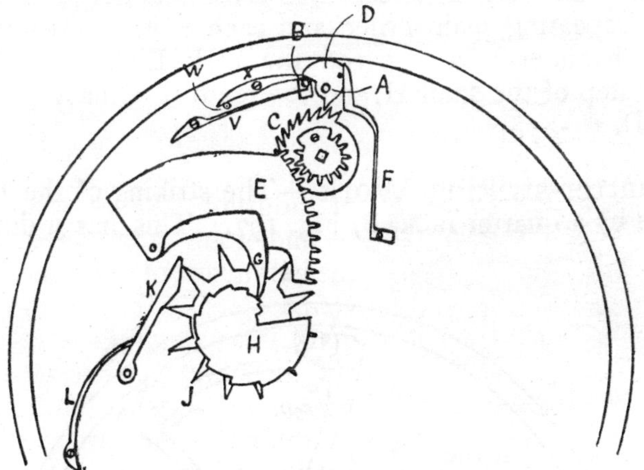

FIG. 196.—Hour-striking Mechanism of Repeater.

pallet; C is wound up by the toothed rack E. In old repeaters E is pushed down by a slide in the pendant, as in Fig. 200. In modern ones E is operated by a slide at the case edge. In some old watches C is wound by a chain passing around it and pulled by a lever operated by a pendant push.

When C is wound up by any of these methods, the teeth pass the point of the pallet D, which trips backwards as the teeth run past it, being returned to its proper position by the light spring F. As soon as it is released, C commences to run forward and the teeth operate on the point of the pallet D, each tooth causing one blow of the hammer upon the gong. The number of blows is thus regulated by the number of teeth of C that pass the pallet. This is determined by the distance to which the rack E can be pushed down. It is pushed down until its lower point G rests upon a step of the hour snail H, and it can be pushed no further. The hour snail is of the same pattern as in clocks, and is mounted upon a "star wheel," J, of twelve teeth, held in position by the jumper K and spring L. It is turned one division each hour by a projecting finger upon the cannon pinion.

Thus, as far as repeating the hours goes, the action is simple. A slide is pushed outside the case; it drives the rack E (Fig. 196) down until its lower point rests upon the snail H. This turns C (Fig. 196) round, a certain number of its teeth passing the pallet D. When released, C returns under the influence of the repeating mainspring, and each tooth causes one blow of the hammer. In Fig. 196 the rack E rests on the ten o'clock step of the snail H, and ten teeth of C have passed the pallet D.

Quarter-striking Work.—The striking of the quarters is done by a quarter rack, P, Fig. 197. This has a double set

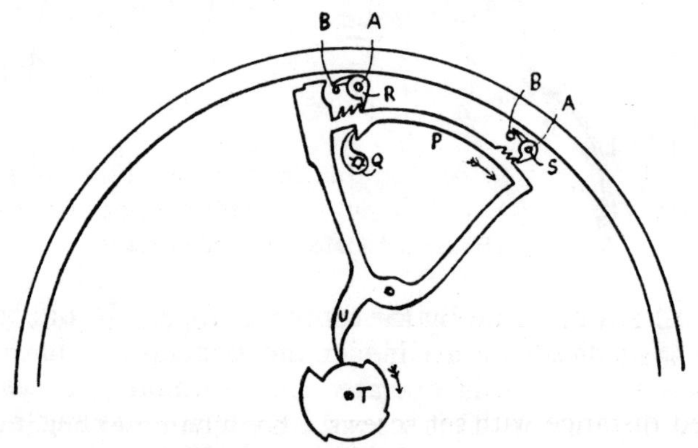

FIG. 197.—Quarter-striking Work of Repeater.

of three teeth, which act alternately on two pallets, R and S, one on each hammer axis.

The pallets R, S, are provided with returning springs, which hold them in position and allow the rack teeth to trip by them as the rack falls. A finger, Q, is mounted upon the squared arbor of the repeating main wheel, and, as soon as the last hour is struck, gathers up the rack P and causes its teeth to act upon the pallets R and S, striking three ting-tang quarters. The distance to which the rack P can fall is regulated by its tail, U, coming into contact with the quarter snail T, which is mounted upon the cannon pinion. The parts shown in Fig. 197 are above and overlay the parts shown in Fig. 196.

Minute Repeating Work.—A minute repeater has, in addition to the quarter rack P (Fig. 197), a minute rack above it, working upon the same stud as a fixed centre. It has fourteen teeth, which actuate a second pallet above the pallet S. Its tail falls upon the minute snail M, Fig. 198, also upon the

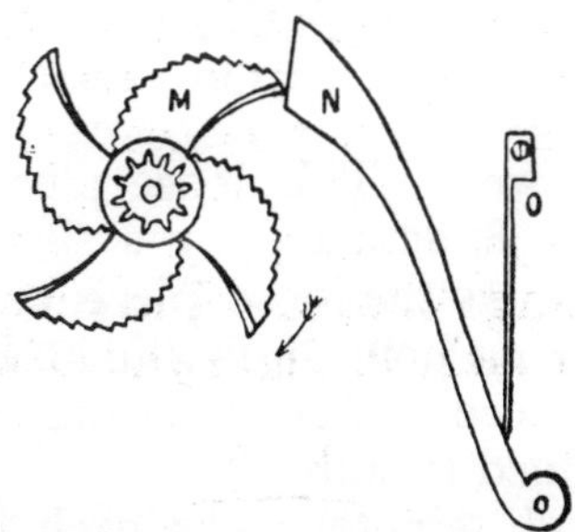

FIG. 198.—Minute Snail and Jumper.

cannon pinion just above the quarter snail. The minute snail has four "wings" each with fourteen steps upon it, regulating the number of minutes to be struck.

Hammers.—The hammers of a repeater do not bank against the gongs, but are held a small distance from them by steel banking pieces (X, X, Fig. 199), which can be set to the required distance with set screws. Each hammer has its spring, which causes the blow. In Fig. 199, X, X are the banking pieces, set by the screws W, W, and V, V are the hammer springs. B, B are the hammer studs, as in Fig. 195. In most repeaters the hammer set screws come through the plate to the back of the watch, but in some old ones they go through the edge of the plate.

Action of Repeating Work.—Just at the hour there is a danger of three-quarters being struck after the hour snail has moved and the next hour has been given. To make this impossible, the quarter snail has underneath it a loose steel plate which, just as the star wheel moves, is flirted forward into such a position that when the quarter rack falls its tail is arrested.

Similarly, the minute snail has a loose plate underneath, having four arms to arrest the tail of the minute rack, so that when the next quarter is struck it will not be possible to strike

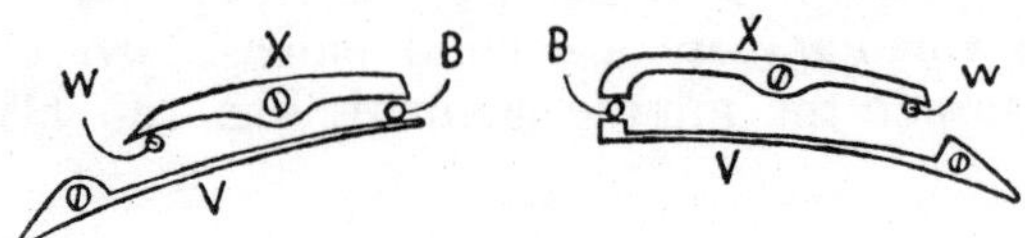

FIG. 199.—Hammer Springs and Stops of Repeater.

fourteen minutes as well. This loose piece is flirted forward by a lever, N, Fig. 198, actuated by a slender spring, O. M is the minute snail. As the loose piece under it passes under N at each quarter of an hour N is lifted a little, and when the edge of the loose piece passes under the point of N it is suddenly flirted or jumped forward.

The quarter and minute racks each have their springs, which, as soon as the racks are released, cause them to fall upon their respective snails. The racks, when arranged as in Fig. 196, are held up by a curved steel spring running round the edge of the movement, called the "all or nothing" piece. Until this piece is lifted nothing can be struck. It is lifted by a small lever just at the moment the rack E (Fig. 196) is driven home against the snail. The exact way this is arranged differs in almost each watch, but its working can be easily seen by inspection. In some E has a loose piece, G, under it, which comes into contact with the snail and lifts the "all or nothing" piece. In others the snail is mounted upon a lever, F, Fig. 200, having a slight movement, and this lever is pressed down when E is driven home, liberating the racks.

Different makes of repeaters vary very much, but the

principles of all are the same. The rack E (Fig. 196) may have an intermediate wheel, B, Fig. 200, between itself and C; or it may be on the right or left of C. The hour rack C is in some old watches between the plates instead of under the pillar plate, in which case the hour hammer pallet is also between the plates. The hammer stops and springs may be arranged in many ways, but the action is always the same.

Fig. 200 shows an older arrangement than Fig. 196. In it

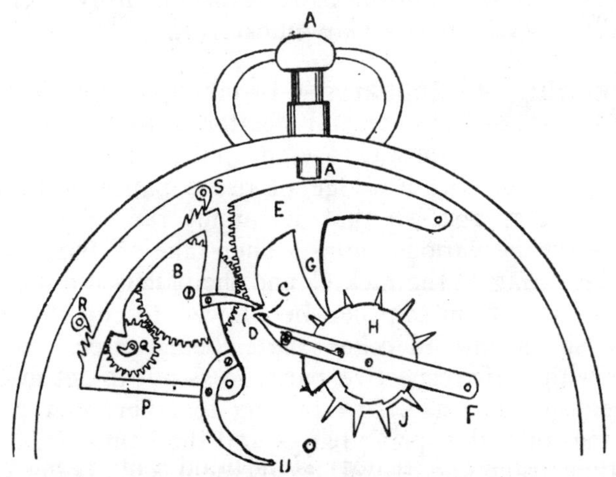

FIG. 200.—Old Form of Repeating Work with Push.

a pendant push, AA, pushes down the rack E until its point, G, rests upon the snail H. H is on a movable arm pivoted at F, and having a very slight motion. This motion is sufficient to depress the point C and liberate the point D of the quarter rack P. The quarter rack then falls until its foot, U, rests on the quarter snail. When the quarter rack is gathered up, the point C of the bar F holds the quarter rack up by the point D. In this arrangement the rack E sometimes drives the hour rack pinion by an intermediate wheel, B, as shown, and sometimes by a chain passing round a pulley wheel or roller. The latter arrangement is a very old one.

Speed of Repeating.—The speed of the repeating train is regulated in old watches by a loose pinion sometimes weighted with a small disc of brass. Its top pivot runs in an eccentric brass bush, which can be turned by a screwdriver to regulate the depth. Shallowing it causes it to run faster. Deepening it makes it slower. More modern ones have a scape wheel and a small pair of weighted pallets, making an escapement which runs through rapidly as the train runs. The speed of these is regulated by limiting the travel of the pallets by a pivoted wire arm, like Fig. 201, which can be turned from the back of the watch by a screwdriver. Giving it less play quickens it.

FIG. 201.—Regulating Curb Pin.

Cleaning a Repeater.—In taking a repeater to pieces, first take off the minute and quarter racks and their springs. Push the slide a little way to wind up the repeating spring, and block the train with a wedge of tissue paper or by the curb shown in Fig. 201, so that it cannot run. This eases the pressure off the various parts. Then take off the rack E (Fig. 196), the spring F, the rack C, and the pallets and their springs. The repeating train can then be allowed to run down. Take off the snails and all other parts, being careful to keep all screws with their respective parts, so as not to get mixed.

Clean all as usual. In putting together, first put in the repeating train, the spring barrel, and the hammers, etc. Then the going train and barrel, escapement, etc., being careful to apply oil to pivots as you go on, as many of them will be covered up by other parts. It is then as well to put the movement in its case, and lay on its back to put on the repeating motion work.

Next oil all the underneath pivot holes. Put on the hour rack C (Fig. 196), the pallets, hammer springs, pallet springs, snail and star wheel and their springs. The square upon which C goes has a dot or a file mark on one side. Let the solid part of the centre pinion of C go next this mark. Block the repeating train. Place a small key upon the square above C, and wind it up about two turns, letting the teeth of C pass the pallet until the last tooth has just passed and the watch would repeat twelve. Then put on the rack E in such a position that it rests upon the twelve o'clock step of the hour snail. The repeating train may now be allowed to run. Put on the quarter and minute racks, their springs, etc., and the finger Q (Fig. 197), pinning it on. Then the "all or nothing piece," and the quarter and minute snails, etc.

The hammer pivots require oil, and their studs where the pallets press them, also the hammer spring ends, the pallets where they turn on the studs and where their springs touch them, the stud upon which the racks turn, the points of contact between all springs and their racks, etc. See particularly that the loose pieces of the snails are free and left quite *clean and dry*, or they will stick.

In cases where there is an intermediate wheel between E and C (as in Fig. 200), the rack, the wheel, and the pinion of C are dotted to show the correct position to put together.

Handwork.—To put the hands of a minute or other repeater on at the correct time, repeat the watch before putting the dial on and note the hour, say seven. Then have a watch going on the board by its side. Watch this as the repeater hands are turned round slowly until the snail moves and changes to eight with a sharp click. Note the time by the other watch. Then put on the dial, seconds and hour hands; put on the minute hand at as many minutes past the hour as the watch travelled since noting it. Finally, test it by holding to the ear as the hands are set and listening for the "click" of the snail moving at the hour.

If the hands of a repeater move very easily they may drag and lag behind when they come to move the hour snail. If not so easy as to drag, but still not very tight, they will, in a minute repeater, be jumped a portion of a minute forward at each quarter of an hour by the lever N (Fig. 198), causing an irregular gain of some minutes per day, which is apt to be very puzzling.

Old Repeaters.—Some old repeaters have the going and repeating trains all between the same plates, and are sometimes awkward to get together. An extremely thin steel hook for pulling and pushing purposes is then useful to get some pivots in. More modern ones are better arranged.

If the repeating train runs much too fast and cannot be regulated, let down the spring a turn or so. Most modern ones have a loose barrel held by a small click. Older ones have to be partly taken to pieces again to do this. If still too fast, as many old repeaters get with wear, weight the fly pinion with a heavier brass disc, a little out of centre, that is, not exactly poised.

Small Faults.—If the hammers jar on the gongs, they are too close, and must be set a little further off. If the gongs foul each other or the case, they can be bent a little, with great care, in the fingers. Sometimes the slide in the case sticks; this is generally through dirt, and if taken out, cleaned, and oiled it will be all right.

A wide bottom centre hole is the cause of many troubles, allowing the quarter and minute snails to be loose and unsteady. Such a hole must be well bushed. Snail loose pieces that stick generally do so through having been oiled. They should be cleaned and left dry. They can be eased by slightly loosening the small collet underneath that holds all together.

In minute repeaters, with wear, the sharp corner gets worn off, N (Fig. 198), causing the minute snail to move too late, and often allowing the minutes to be struck after the hour has changed. Also, this fault often stops the watch by allowing the minute rack to fall, when there will be no time for it to be gathered up before it is overtaken and jammed by the arm of the minute snail. The remedy is to re-point up N centrally, by careful grinding with oilstone dust on a flat steel polisher and polishing with red-stuff. Just grease the point of N very slightly. The exact time of flirting of the minute snail may be adjusted by working the point of N backward or forward by grinding the forward or backward slopes as required. If it flirts too soon, the period during which fourteen minutes are struck will be a very short one, or may be absent altogether. If too late, the rack may jam as described.

A complete failure of one hammer to strike is generally caused by its pallet sticking. See that all are perfectly free and easy, and that their recovering springs act promptly.

A failure to strike the last minute or stroke of the quarters is often caused by the slide sticking when nearly home. Dirt accumulates at the end of its slot and prevents it from going quite home.

Clock Watches are on a different system to repeaters. They have two trains and mainsprings, both wound by the same keyless work, forward winding one spring, backward winding the other. As the watch strikes continuously, they both run down together. To repeat these watches, a slide is moved a little only. It operates a small lever inside and releases the striking train, which then runs and repeats the last hour, etc. The slide does not wind up a spring, as in a repeater, but only liberates a small catch.

Clock watches are very complicated pieces of mechanism, and look to the uninitiated a mass of springs, levers, and racks. But to a man who has thoroughly mastered minute repeating work they will present no difficulty. The great thing in cleaning them is to keep each spring with its lever or rack and all screws with their pieces; to pin on and oil each part as it is put on, and see that all are free, and to oil all contacts of parts with their springs; to look out for dots on wheels and racks, etc., and be careful of them; to leave all flirting pieces *dry*, so that they cannot be fixed by oil. This is especially true of the mass or bundle of pinions, etc., on the squared axis of the striking wheel. There are several flirting actions there, and oil will be a fruitful cause of trouble.

There have been other forms of both repeaters and clock watches made, but by far the greater number are as here described. One form of repeater deserves some notice. It has the hour-striking teeth and the quarter teeth all on one steel wheel, and the pallets have a "rise and fall" motion, putting them into and out of gear with the teeth. A careful inspection of the parts before taking apart will be sufficient to show its action and be a guide to putting it together again.

Some very old watches—generally verges—have bells instead of gongs. These, of course, take up much more space, and in modern watches would not be tolerated. Others were "dumb repeaters." In these the hammers struck steel blocks in the case, causing a dull thud only, which could be felt by the hand but hardly heard at all.

Alarm Watches have been made which could be set to go off at any hour, much in the same way as a modern alarm clock, and generally were operated by very similar means. They will offer no special difficulties. The same remarks apply to **Musical Watches**, which play a tune at the hour or when required. Such watches are out of date, and the "music" generally of a poor quality. In them there is usually a stack of flat steel gongs, operated by a pin barrel like a musical box, and a train of running wheels governed by a fly pinion.

Broken Parts.—Any of these watches can be repaired by a workman who has learned to file and turn properly. Broken repeating parts may be sometimes repaired by brazing with brass spelter, re-hardening, and tempering. Some parts on which there is little strain may be soft soldered. New parts can be made by filing and turning from rough steel, being hardened, tempered, and smoothed off to shape. Teeth may be dovetailed and soft soldered in racks. The pivots upon which most parts work are steel studs screwed into the plates, or screwed into the parts themselves, and the sockets which work on the studs are also screwed into the racks, etc. Hammer arbors are screwed in to a shoulder. Hammer studs are also screwed.

A new minute hand to a repeater with a square cannon pinion is difficult to fit, on account of filing the square to exactly face the right way to correspond with the striking. Hands for these watches can be bought, with revolving sockets. When filed out to fit, the hand can be turned round on its socket until correct, and then riveted up tight.

Old English chain repeaters sometimes break their chains. In mending, new links must be supplied or made of the *exact length* of the old ones, or difficulties will arise and wrong numbers will be struck.

The general remarks on p. 208 apply equally to repeaters, and those not thoroughly familiar with the work are advised to study them.

CHAPTER XIX

CHRONOGRAPHS, CALENDARS, ETC.

Stop Watches.—These are merely centre seconds watches with a slide at the side of the case, operating a lever on the pillar plate. The lever has an upright thin brass wire, which comes into contact with the balance rim or the roller edge and stops the watch. Pulling the slide back releases the balance, and the watch starts again. An improved form of stop watch has the stopping mechanism operated by pressing the keyless winding button.

The **Independent Centre Seconds** watch was introduced to enable the seconds hand to be stopped and restarted without stopping the watch at all. These watches had two mainsprings and two trains of wheels. One was the going train, the other the centre seconds train, and its sole business was to drive the seconds hand. The scape pinion had always six leaves, therefore one leaf came round per second. The last pinion of the independent train was a fly pinion, with a light arm that just caught the leaves of the scape pinion. Every time a leaf passed, the arm of the fly pinion was allowed to pass, and the fly made one revolution. This caused the centre seconds hand to beat dead seconds. A few showed fifths of seconds by a small hand making one revolution per second. The independent train was stopped and started by a slide at the case side, as in stop watches. The slide worked a lever which caught the fly pinion arm. Independent centre seconds watches are no longer made, their place being taken by chronographs with a "start stop and fly back" action quite independent of the going train.

Centre seconds watches have the minute hand driven by

motion wheels. If the depths of these wheels are a little shallow, they allow the minute hand too much shake. Very often the amount of shake is equal to as much as one minute division on the dial. This renders the time uncertain to that amount. To reduce the error, some watches have a very thin and light spring acting on the teeth of the cannon pinion to hold it back, as in Fig. 202. Such a spring can with advantage be fixed to

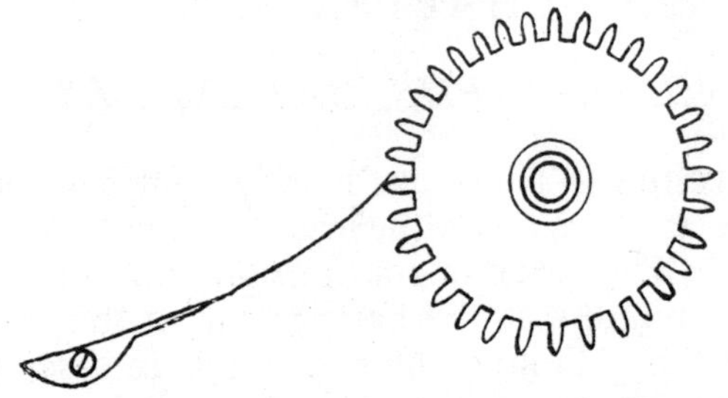

Fig. 202.—Chronograph Motion Work Spring.

a good watch of this kind having shaky motion work. Its back pressure is so slight that it does not interfere at all with the going of the watch.

Chronographs.—The first chronographs really *marked* the time. A seconds hand was used whose point held a small drop of ink. On pressing the button, this was caused to mark the dial with a dot without stopping the hand. Modern chronographs start the hand with the first push of the button, stop it by the second, and cause it to fly back to zero by a third push. When not in use as a chronograph the hand rests at zero.

Starting Mechanism.—They are arranged in many ways, but all contain the same general mechanism and act on the same principles. A centre seconds hand is mounted on an axis with a fine-toothed wheel. Another wheel the same size is mounted on the fourth pinion of the going train, or the fourth wheel itself used. An intermediate wheel between the two is so arranged that it can be put into or out of gear by a slight movement. When in gear the centre seconds hand runs. When out of gear the hand stands still.

Fig. 203 is an illustration of one method adopted, in which the fourth wheel itself drives the small intermediate wheel A. A has two sets of teeth. One set gears with the fourth wheel

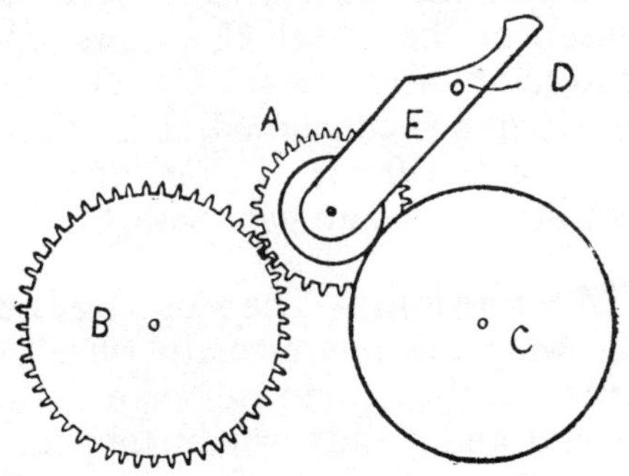

FIG. 203.—Driving Arrangement of Chronograph.

B, and the other, very fine set, gears with the centre seconds wheel C. A is mounted in a frame, E, turning on a pivot at D. It has a very slight movement, just enough to engage and

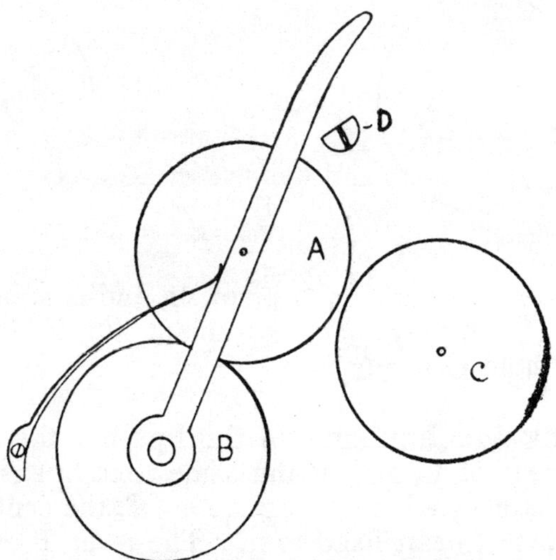

FIG. 204.—Driving Seconds Hand of a Chronograph.

disengage the fine teeth. B is the fourth wheel, with which A is always running; C is the centre seconds wheel. In Fig. 204, B is a fine wheel mounted on the fourth pinion; C is the centre seconds wheel; A is the intermediate wheel running in a frame like a lever, pivoted at the fourth pinion. A very slight movement of the lever D suffices to throw A into gear with C. In other watches the wheels have bevel teeth, and the intermediate wheel, or the wheel C, throws into and out of gear by a rise and fall motion.

So far as the starting is concerned, it is effected by a press of the button moving the frame E (Fig. 203) or the lever D (Fig. 204) so as to throw A into gear with C.

Stopping Mechanism.—The stopping is effected by the second push throwing the intermediate wheel A out of gear again, and simultaneously applying a brake to the rim or teeth of C to hold it still and steady where stopped. In Fig. 205,

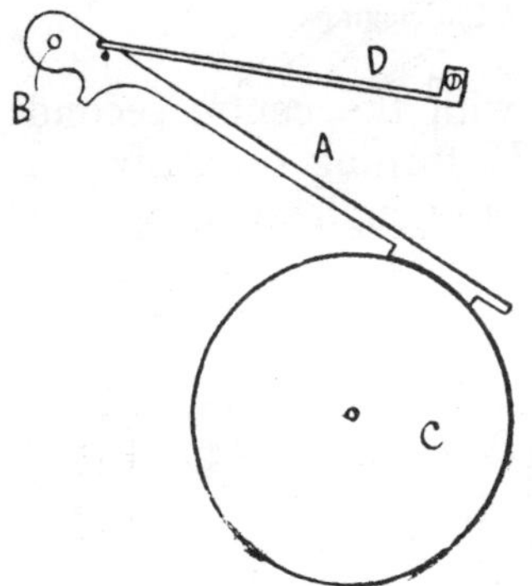

FIG. 205.—Brake.

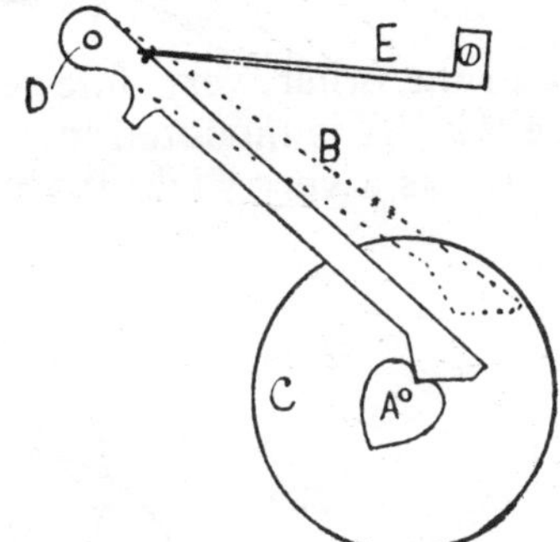

FIG. 206.—Fly-back Work of Chronograph.

A is the brake. It works on a pivot, B, and is applied by the light spring D. A very slight movement lifts it clear of C or applies it to hold C steady.

Fly-back Mechanism.—A third push of the button lifts the brake clear of C, and at the same time brings C back to zero. This is effected as in Fig. 206. C, the centre seconds wheel, has a steel heart fixed to it. The lever B comes down sharply upon the heart, which is really a part of a spiral, and by its pressure on the polished edge of the heart causes C to fly round until B rests upon the hollow and holds C steady at zero. B is pivoted at D, and is brought down sharply by a spring, E

The fourth push of the button commences the cycle of operations again, and lifts B (Fig. 206) clear of A, and simultaneously throws A (Figs. 203 and 204) into gear with C.

Driving Work.—Fig. 207 shows how all the mechanism is worked by pressing the button.

The button, upon its being pushed in, depresses the lever A, working on a stud or pivot at B. At its other end is a hook, C, which engages with the ratchet wheel D. Each time A is depressed D is hooked round one tooth. D is held in position

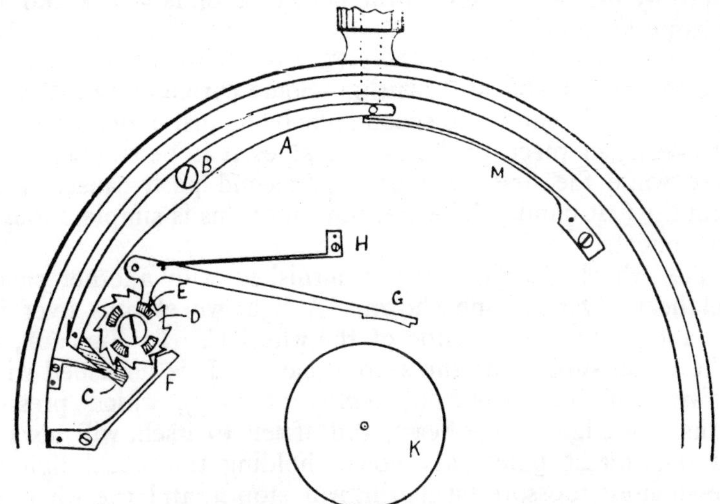

FIG. 207.—Driving Mechanism of Chronograph.

by the click F. Upon D are four projecting pieces of steel, E. It takes three pushes of A to bring each block E round, forming a complete cycle. Against the blocks E rest the tail ends of the levers applying the brake, throwing the intermediate wheel into gear and bringing the heart to zero. D is therefore the main controlling wheel of the entire mechanism. Each lever has its spring, holding its tail end against the blocks E. As a block comes round it raises one tail end, as at G, and then allows it to fall into the space again.

Minute Recorders.—Most chronographs are furnished with a minute recorder.

The wheel C or A (Fig. 204) is fitted with a short finger, which comes round at the end of one minute and moves the minute recording wheel round one tooth. A hand on the dial indicates the number of minutes run since the hand was started. Generally the recording hand reads up to thirty minutes, and therefore the wheel has thirty teeth. Minute recording mechanism varies very much in design, but is always fairly simple and easily seen. Some minute recording hands are slowly travelling all the time the hand is running. The minute recording wheel is furnished with a heart piece, and is brought to zero by the same lever as the centre seconds wheel, and in the same manner.

Double Fly-back Hands.—Some chronographs have a double seconds hand, the second one being just underneath the first. A push piece in the case enables it to be stopped anywhere while the first runs on. A second push causes it to rejoin the first hand. The mechanism of this is simple, though very delicate.

The wheel C (Fig. 206) is furnished with another small steel heart over the one shown. A light wheel runs over it, with a pipe outside the pipe of the wheel C. In Fig. 208, A is the wheel supporting the second hand. It is furnished with a pivoted friction roller (generally of ruby), which presses against the edge of the heart, and, if left to itself, will always find the lowest point, as shown, holding the wheel lightly. The slightest pressure on its rim will stop it, and the wheel C (Fig. 206) underneath it, with the heart, will continue to travel without it. The small ruby roller meanwhile runs round the heart. A (Fig. 208) has a smooth rim with no teeth.

The stopping mechanism for the wheel A consists of a double brake like a pair of scissors, as at B. These are allowed to close a little by a spring and grip the rim of A. This stops the hand. When B is opened the wheel A is released, and the small ruby roller and spring, acting on the heart on C, causes A to return to its original position on C and travel with it again.

The brakes B are alternately opened and closed by a square between their outer ends. When two angles are presented to B, they open; when two flats, they close. They are closed by a spring. The square is worked round by a ratchet and a hooked lever actuated by a case push piece, as shown in Fig. 208, very much resembling the mechanism shown in Fig. 207.

Cleaning.—To clean a chronograph, take out the balance for safety; then remove the chronograph work, and take all to pieces, being careful to keep each spring with its lever, etc., to avoid confusion. In putting together, put on the chronograph work last, preferably when the watch is in its case. Oil the pivot of the wheel D (Fig. 207), all the lever pivots, and the blocks E. Oil where each spring makes contact with its lever.

When the chronograph hand is mounted on a long, slender pivot coming through the centre pinion, and its wheel is on

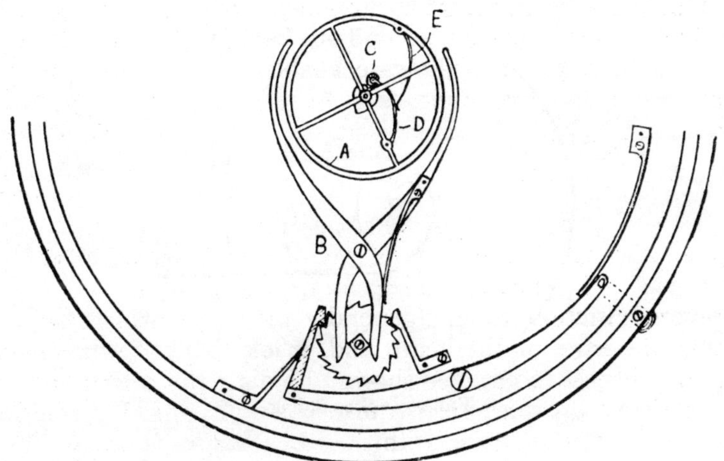

FIG. 208.—Arrangement of Double Fly-back Chronograph.

the back of the watch, oil its pivots. But when the hand is mounted on the pipe of its wheel, which runs outside the cannon pinion or centre arbor, the pipe must be absolutely clean inside, and not oiled. The slightest oil or dirt here will stop it. These remarks apply still more forcibly to the second hand of a double fly back. Its pipe and ruby roller must be clean and free. Some chronograph hands do not always go back exactly to zero, but vary about $\frac{1}{5}$ second each time. To remedy this, many are fitted with a light spring, which just catches the point of the heart and holds it exactly true each time. While the hand is running the point of the heart just touches the tip of this spring at each revolution and passes it. Fig. 209 shows the arrangement. A is the spring that holds the heart-point.

The fine-toothed wheels are very delicate. They must be brushed clean in the teeth, as the slightest dirt completely fills up the fine-cut spaces. The depth between A and C (Figs. 203 and 204) is a very particular one. In some watches A is simply kept pressed into C by a light spring, and the depth has no shake at all, one wheel following the inequalities of the other. In these watches starting the seconds hand nearly always takes off some of the power, and causes the action of the balance to fall off very much. In other watches the lever carrying A is pressed against an adjustable stop (D, Fig. 204),

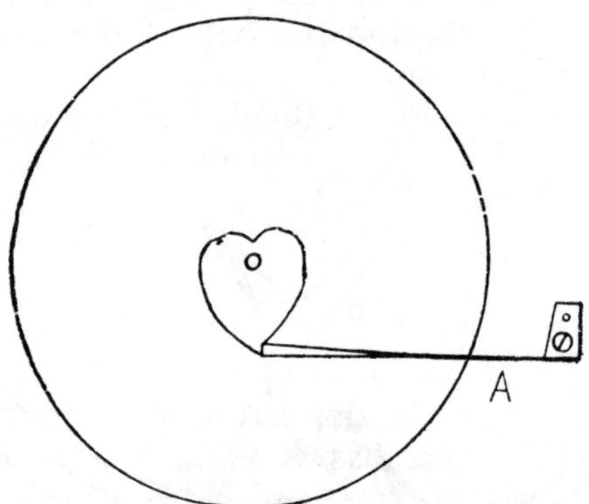

FIG. 209.—Stop Spring.

and can approach C no further. In these the depth can be set to a nicety with a strong eyeglass, so as to have just a little shake all round. It causes the point of the hand to move a little loosely, but avoids dragging the watch.

Double fly-back chronographs are very delicate, and seldom work perfectly for many months together. The slightest dirt will cause sticking of the hands, and prevent them flying back properly to zero. Great care must always be taken in putting the hands on to see that they are all free of each other in any position they may take up.

There is one form of Swiss chronograph often met with in which the chronograph hand and dial are both at the back of the watch. In these the chronograph wheel C (Fig. 204) is driven direct by the wheel B, on the fourth wheel axis, and there is no need for an intermediate wheel. Both these wheels, B and C, are bevel wheels, and C throws into gear by a rise and fall or "pump" action. They are so arranged that by turning two or three dog screws at the edge, the whole of the chronograph work can be removed bodily without disturbing its parts, and, if necessary, the going part of the watch can be cleaned and repaired without doing the chronograph work.

Simple Calendars.—In these the day of the month is shown, and the watch automatically changes it at midnight each 24 hours. The date dial is divided up to 31 days, and for the months having that number of days is correct. For the 28, 29, or 30 day months the hand requires correction. Some calendars show in addition the day of the week, the month, and the age of the moon.

In these simple calendars the day of the week, the month, and the moon dials will need no alteration so long as the watch is wound (except for a small error that accumulates with the moon disc), the date only requiring correction.

The calendar wheels are driven from the hour wheel, which turns once in 12 hours. This is made to drive two wheels of exactly twice the size, turning in 24 hours. A pin, B, in one of these wheels moves the day of the week wheel, C (Fig. 210), one tooth. This wheel has 7 teeth, and is mounted on a fixed stud in the watch plate, and held in position by a spring or jumper, A. The other 24-hour wheel similarly moves the date wheel, made and mounted as in Fig. 210, but with 31 teeth. A pin or finger on the date wheel moves the month wheel (12 teeth) one tooth at each revolution. A second pin on one of the 24-hour wheels moves the moon wheel one tooth per day.

In cleaning these watches, leave all calendar wheels and studs dry, but oil the springs where they touch. When putting together, set the watch at 12 o'clock; then put on all calendar wheels in such a position that they just begin to change. As a rule, each hand and wheel can be set by jumping it round with a push in the case side. Take out and clean all these pushes, as dirt is apt to accumulate and fix them.

Perpetual Calendars never want altering, unless the

watch stops. They show the date correctly, and change the month in February, in 30 and 31 day months, and even give February 29 days in leap year. But as they all failed in the year 1900, they can scarcely be called perpetual! Still, it will be many a year before they need go wrong in this way again.

As before remarked, it is the date wheel that causes the difficulty. Instead of being moved by a simple pin set in a 24-hour wheel, it has to be moved by a jumper. This is a steel lever that is drawn back a certain distance, and in going

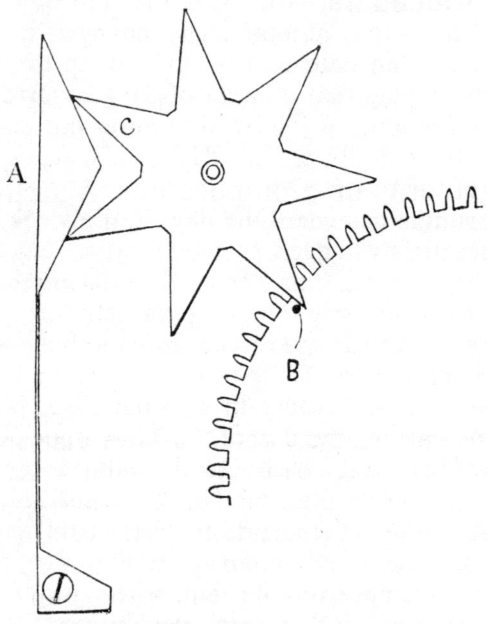

FIG. 210.—Arrangement of Calendar Work Jumper.

forward pushes the date wheel on one tooth. An arrangement is made by which, at the end of April, June, September, and November, the jumper is drawn back more, and in going forward moves the wheel *two* teeth. In February it moves it *four* teeth, except in leap year, when it only moves *three*.

There are many different ways of effecting it. In some there is a wheel of 48 teeth, moving one tooth per month and making a revolution in four years. It has shallow slots in its rim for the 30-day months, a deeper one for one February of

29 days, and three still deeper ones for February of 28 days. These slots regulate the distance the jumper is drawn back. There are almost as many ways of making perpetual calendar work as there are calendar watches, and any illustrations would be useless.

It is advisable to thoroughly study the action before taking any of it apart, and as a rule little difficulty will be experienced. Take care to leave it on the right month and the right year; also to see that when the month changes the date hand drops on to the "1st" exactly.

In nearly all calendars the work is mounted entirely upon a loose frame, put on over the motion work and held by two or three screws at its edge. This enables it to be all removed together, and replaced last thing in putting the watch together.

Arrangement of Complicated Watches.—Many watches show nearly everything that can be shown and do everything that can be done. The writer has cleaned watches that are (1) clock watches, (2) minute repeaters, (3) double fly-back chronographs, (4) perpetual calendars, and (5) aneroid barometers, all in one. In such watches the pillar plate is very deep and hollow. Next the plate comes the clock watch and repeating mechanism. Above that, supported by bridges and cocks, lies the chronograph work. Above that, in its separate frame screwed on over all, is the calendar work. The barometer has a circular vacuum box in the shape of a ring, and it lies round the movement outside the repeater gongs.

Occasionally a repeater and perpetual calendar has the repeating work on the back of the movement and the calendar work under the dial. Many Swiss chronographs and calendars have the chronograph work at the back and the calendar work under the dial.

Double-time Watches are not very complicated, merely having a double set of motion work and showing two times—say, Greenwich and Paris—on two different circles.

Sidereal Watches have a second pair of hands showing sidereal time. Neither of these last require any special directions. The latter depend on the ratios of the wheels employed, and not on the putting together.

General Remarks.—The cleaning and repairing of complicated watches is not work that should be attempted by a novice.

Careless handling is the cause of innumerable troubles. In putting parts together, if something does not go quite right, an inexperienced workman is tempted to alter this or that piece, to "just take a little off" here or there with a file, in the belief that all will then be right. In such a case it is well to consider first that the man who made the parts in question undoubtedly made them right, and therefore, if the action is not right afterwards, it is either that the mechanism is not correctly put together or suffers from the effects of wear. It is, in fact, much more likely that some parts require a little putting on than that they want any filing off.

A little filed off one part, it will be found, necessitates a little being filed off another, and so on, until the entire action is almost ruined beyond repair.

Therefore, unless a workman is thoroughly at home with this kind of work, or is merely practising on an old and worthless movement, he should not attempt on his own responsibility to alter the form of any part of a complicated watch.

INDEX.

ACHROMATIC eye-glass, 27
  Acids, 16
Adjusting rod, 184
  for temperature, 131, 140
Alarm watches, 195
Aluminium bronze, 13
American watch lathe, 41
Arbors, set hand, 151
  turning, 37
Arcs, long and short, 141

BALANCE, 129
  compensation, 130
  poising, 111
  raising, 113
  staff, turning, 110
Banking pins, 106
Barrel, closing, 74
  cover, tightening, 74
  hook, to put in, 75
  new, fitting, 75
  raising or lowering, 76
  teeth, replacing, 76
Bent teeth, straightening, 76
Benzine, 15
Blue on steel, removing, 93
Blueing slip, 52
  screws, 99
Boley millimetre gauge, 25
Bows, 159
  keyless, fitting, 159
Brass, hardening, 11
  properties of, 10
  softening, 11
  tempering, 11
Breguet springs, 137
Broaches, 24
Broaching, 34
Bruises in bezels, 157
  in cases, 158
Bushing pivot holes, 90

CALENDARS, perpetual, 205
  simple, 205
Cannon pinions, fitting, 148
  tightening, 149
Cap studs, fitting, 98
Carrier for lathe, 44
Case joints, 158
Cases, 155
Cementing wheels in lathe, 85
Centre seconds motion wheels, 198
Chain, to mend, 77
  hook hole, to make, 75
  putting on, 64
Chalk, 15
Chamfering tools, to make, 51
Chronographs, 198
  cleaning, 203
Chronometer escapement, 125
  balance screwdriver, 184
  marine, 182
  new pivots in, 184
  pocket, 67
  stopping, 174
Chuck, split wire, 45
  wheel, 45
Chucking wheels, 85
Cleaning duplex watches, 66
  English cylinders, 66

Cleaning gold hands, 68
 jewelling, 69
 mainsprings, 57
 nickel movements, 68
 pivot holes, 58
 pocket chronometers, 67
 tarnished gilding, 68
 verge watches, 66
 watches, 54
 watches, general remarks on, 67
Click springs, fitting, 81
Clicks, fitting, 81
Clickwork, 3
 repairing, 80
Clock watches, 194
Club-roller, 102
Collet, hairspring, 133
Compensation balance, 130
 curbs, 147
 trueing, 131
Complicated watches, 207, 208
Cone cement chuck, 89
 pivots, turning, 111
Conversions, 176
Countersinks, to make, 50
Crocus, 13
Curb pins, to replace, 147
Cutters for jewelling, 97
Cutting nippers, 22
Cylinder banking pin, 117
 depth, to alter, 115
 English, 120
 English, cleaning, 66
 escapement, 114
 height, 116
 plugs, 119
 turning, 117

DART, to advance, 104
 Depths, 82
Detached lever, 102
Detent, 125
 to make, 127
Dials, enamel, 152
 metal, 153
 pins to remove, 62
Diamantine, 14
Diamond powder, 14
Dog-leg lever, 179
Domes, raising, 160
Double fly-back chronographs, 202
Double roller, 103
 time watches, 207
Douzième gauge, 25
Dovetail slips, fitting, 95
Draw, testing, 109
Drawing a pivot hole, 83
Drill holders, 45
Drilling, 33
 pinions, 93
Drills, making, 48
 ready made, 23
 sharpening, 34
 tempering, 10
 very small, hardening, 10
Dumb repeaters, 195
Duplex escapement, 123
 watches, cleaning, 66

EMERY, 14
 Endshake of wheel, 59
Endstone slips, fitting, 95
 fitting, 95
Escapement, faults in, 171
 use of, 5
Eye-glasses, 26
 achromatic, 27
 double, 27
 holding, 28
 steaming, 28

FILES, 23
 use of, 30, 32
Filing flat, 30
 pins, 31
 tempered steel, 33
Finger marks, to remove from dials, 153
 plates, 69
Flat filing, 30
Flat spring, altering to breguet, 180
Flatted ruby pins, making, 107
Fly-back mechanism, 200
Fly springs, 158
Follower, 66
Free springs, 143
Fusee click, to make, 78
 keyless work, 168

Fusee, pinning up, 79
ratchet fitting, 78
re-cutting, 77
stop, correcting, 79
to take apart, 77
top hole, putting in, 91
turning arbor, 78
use of, 3

GAUGE, douzième, 25
mainspring, 73, 74
pinion, 25
sliding vernier, 25
Geneva stop-work, 57
German pattern watch lathe, 42
German silver, 12
Gilding, tarnished, to clean, 68
Glasses, fitting, 157
forms of, 157
Going barrel, 3
Gold, use of, in watches, 12
Gold-filled cases, 156
hands, cleaning, 68
Graver, how to hold, 36
points, 37
Guard depth, correcting, 105

HAIRSPRINGS, 132
collet, 133
counting vibrations of, 134
forms of, 138
gold, 12
pinned in at equal turns, 140, 142
repairing, 145
selecting, 133
to pin in, 135
to set flat, 136, 137
use of, 5
Hammers, 23
Hand tongs, 91
Hands, to tighten, 149
fitting, 150
Hard soldering, 52
Hardening brass, 11
steel, 9
very small drills, 10
Heart piece, 200
Heating workshop, 19
Hollow fusee, 91
Horizontal escapement, 114
Hour wheel, to tighten, 150

INDEPENDENT centre seconds, 197
Invar, 13, 145
Isochronism, 141

JEWEL holes, 5
Jewelling, 96, 181
Jewelling cutters, 97
to clean, 69
Joint pins, to remove, 62
pusher, to make, 51

KARRUSEL watches, 145, 175
Keyless conversions, 180
Keyless watches, American, 67
English, 67
removing from cases, 61
Keyless work, 161
fusee, 168
Patek Phillippe, 166
pendant-set, 166
rocking-bar, 161
shifting sleeve, 164

LATHE, 40
American pattern, 41
carriers, 44
driving, 46
German pattern, 42
Lever, dog-leg, 179
Lever escapement, 100
making, 176
pitching, 178
setting in angle, 178
straight line, 101
Lever, to remove, 63
watch, to set in beat, 65
Lever rack, 102
Lever-set work, 162
Levers, making, 177
Lighting the workshop, 19
Locking stone, 127
Loose pulley runner, 44

MAGNETISM, protection from, 174
to detect, 174
to get rid of, 174
Mainsprings, 2
bad, 175
braces for, 72
breaking, 74
cleaning, 57
fitting, 70
gauges for, 74
hooking, 71
letting down, 55, 61, 63
number of turns of, 2
setting up, 57, 64
strength gauge for, 72
winding in, 73
Maintaining detent, correcting, 79
work, easing, 80
Mandrel, 42
Marine chronometers, adjusting, 183
cleaning, 182
Mercury, 16
removing from plates, 16
Methylated spirit, 15
Millimetre as a measure, 25
Minute recorders, 201
Mis-locking of levers, 104, 105
Motion work, 6, 148
trains, 149
Movement, 1
taking out of its case, 54
Musical watches, 195

NAMES, removing from dials, 153
Nickel movements, 12
to clean, 68

OIL for watches, 15
Oiler, 52
Oiling pivots, 59, 60
Oilstone-dust, 14
Oven, 141
Overcoils, forming, 139
forms of, 142

PALLADIUM, 13
balances, 13
springs, 13
Pallet depth, testing, 105
stones, cementing, 109
Pallets, correcting, 108
Patek Phillippe keyless work,
Pegwood, 16
Pendant-set keyless work, 166
Petrol, 15
Pillars, split, 97
Pin filing, 31
Pin vice, 25
Pinion gauge, 25
Pinions, drilling, 93
facing, 87
fitting, 87
Pin-pallet escapement, 101
Pins, to handle, 69
Pith, 17
Pivot holes, bushing, 90
centring in lathe, 97
cleaning, 58
ends, rounding up, 94
polisher, 86
uprighting, 96
Pivots, polishing, 84
putting in new, 92
short, 174
straightening, 112
Platinum, 13
Play of hairspring, 147
Pliers, 21
brass-nosed, 21
small, making, 21
Polishing pivots, 84
rollers, 178
Position errors, 141
Progress screws, 25, 98
Punches, 24
making, 49
Push pieces, 158
Putting a watch together, 59

QUARTER repeater, 188
screws, 131

RATING watches, 141
Red stuff, 13

Repeater hammers, 187
hands, 193, 196
Repeaters, 186
cleaning, 192
Repeating hours, 187
minutes, 189
quarters, 188
speed of, 191, 193
spring, 186
work, faults in, 193
Rocking-bar keyless work, 161
butting, 163
Rollers, club, 102
making, 177
polishing, 178
removing, 108
Rouge, 13
Rounding up pivots, 94
tool, 83
Ruby pins, 107
cementing, 107
holes, making, 177
roller, 123
Running turning centres, 90

SAFETY action, 103
barrels, 80
pinions, 80
turning centres, 94
Scape wheels, riveting on, 109
Screw cases, 156
Screwdrivers, 22
small, making, 50
Screws, fitting, 98
broken, to remove, 98
re-blueing, 181
threads, to cut, 35
Seconds pieces, cementing in, 152
Set hands arbors, 151
Setting in beat, 65
Sharpening drills, 34
Shellac, 17
Shifting sleeve keyless work, 164
Sidereal watches, 207
Silver, 12
paper, 17
Slide rest, 43
Sliding tongs, 22
Snaps, tightening, 156
Soft soldering, 50
Softening brass, 11
steel, 8
Soldering, 50, 52
Spirit lamp, 15
Spirits of salt, use of, 16
Split chuck, 45
Square shoulders, turning, 88
Stakes, 24
Steel, hardening, 9
maximum elasticity of, 9
properties of, 8
softening, 8
tempering, 9
Step chuck, 45
Stop watches, 197
Stopping watches, 170
Stopwork, Geneva, 57
Straight line lever, 101
pivots, turning, 88

TAPPING holes, 35
Taps, to make, 48
Teeth, bent, to straighten, 76
broken, replacing, 92
Temperature, adjusting for, 140, 144
error, 129
Tempered steel, to file, 33
Tempering brass, 11
steel, 9
Three-quarter-plates, improving, 181
watches, 65
Tightening hands, 149
Timing in positions, 141
watches, 140
Tissue paper, 17
Tools, making small, 48
Train, finding stoppages in, 170
wheels, 4
Trains, motion work, 149
to count, 133
Turning, 36
how to practise, 40
with a lathe, 40
Turns, 36
Turpentine, 16
Tweezers, 22, 53
Two-pin escapement, 104

UNEQUAL rates, 175
Up and down work, 154
Uprighting pivot holes, 96

VERGE conversions, 179
escapement, 121
holes, putting in, 122
watches, cleaning, 66
Vices, 20

WATCH movement holder, 58
Water of Ayr stone, 15
Wear, causes of, 84
Wheel chuck, 45
cutting frame, 43
Wheels, riveting on pinions, 88
trueing, 84
uprighting, 96
Winding wheels, teeth of, 168
Wire chuck, 45
Work-board, 20
Workshop, heating and lighting, 19

THE END.